A MODERN MASON EXAMINES HIS CRAFT: FACT vs. FICTION

A MODERN MASON EXAMINES HIS CRAFT:

FACT vs. FICTION

by L. C. Helms

MACOY PUBLISHING & MASONIC SUPPLY CO., INC.
Richmond, Virginia

L.C. Catalog Card No. 81-84637

Printed in the United States of America

Author's Preface

A Modern Mason Examines His Craft: Fact Versus Fiction is the result of this writer attempting to balance the facts and fiction that appear to surround Freemasonry. It is not intended to serve as a guide to Freemasonry for the non-Mason, although it could serve that purpose, but it is directed to the Freemason who is curious about the history of certain practices currently utilized in Masonic lodges. This writer has grown increasingly concerned about the confusion and misinformation associated with the Craft—misinformation that is prevalent even amongst tenured Freemasons.

Why was Freemasonry, for example, considered by many individuals to be anti-Catholic or vice-versa? Why is there such an emphasis on secrecy in a day-and-age when every aspect of the Craft's ceremonies can be obtained by the curious non-Mason in one hundred different books? Why do men join the various Masonic orders today as opposed to the reasons they joined in the past centuries? How old is the Institution and what are its historical roots? How should the Masonic allegories be interpreted?

These are among the confusing items to many Freemasons irrespective of their previous instructions. There appear to be misunderstandings in separating the historical truths of the origins and aims of Freemasonry and the web of fiction that surrounds it. Although many men attend Lodge faithfully and/or serve as officers and thereby become more readily conversant with Freemasonry, the candidate who passes through the Blue Lodge degrees is often nervous and sometimes in the dark about many aspects of the ceremonies. Even when receiving lectures, the candidate is often tense and his comprehension and retention are diminshed. This short work, therefore, is designed to clarify certain aspects of Freemasonry, to make the more salient aspects of the Craft more easily understood. It is an attempt to create a primer; where various pieces of information can be brought together under one cover, that will serve as an easy reference and aid. Obviously, this work will cover only this writer's perceptions. It is not intended to be encyclopedic in nature, but rather selective in the topics covered. In presenting this material I am reminded of the words of Albert Pike in his classic *Morals and Dogma:*

> "In preparing this work...[I] have been about equally author and compiler; since [I] have extracted half of its contents from the works of the best writers and most philosophers or eloquent thinkers. Perhaps it would have been better, and more acceptable, if [I] had extracted more and written less." [1]

Acknowledgments

I would like to thank my brethren of Tanana Lodge No. 162 F. & A.M., Fairbanks, Alaska and Ellensburg Lodge No. 39, F. & A.M., Ellensburg, Washington for their guidance and support in my Masonic life. Although there are countless individuals who have contributed directly or indirectly to this work, I especially acknowledge and thank Harry Sharpe and Harleigh Lines of Yakima, Washington, Charles Lloyd of Fairbanks, Alaska, and Hiram Colburn, James Richard DeVries and Eric Brattkus of Ellensburg, Washington.

TABLE OF CONTENTS

I

Introduction

It is asserted by Freemasons that Masonry is not a secret society, but rather a society of secrets. To most non-Masons this equivocation is simply a semantic exercise designed to further camouflage the Society and hide nefarious goals and objectives. The fact that any part of Masonry contains secrets or alludes to esoteric knowledge assists the damning mind into drawing unwarranted conclusions on facts certainly not in evidence. These phenomena are not restricted to Freemasonry or anti-Masonic activities, but are corollated to the entire spectrum of secret societies or bodies that hold esoteric knowledge. But, to focus directly on Freemasonry:

> "Let us consider Freemasonry. Anyone who lives in Britain or the United States is certain to have Freemasons among his acquaintances; there may be a Masonic Temple in the center of town, or it may be known that a local Masonic lodge meets on given nights at a particular hotel. Nonmembers may attend dances or dinners organized by Masonic lodges, or hear of their charitable activities. Yet, Masons learn to preserve the secrets of their Craft, and outsiders know little of Masonic beliefs and ritual. What attracts a man to join such a group? It is obviously not enough to dismiss such going-ons as a childish liking for dressing-up, as some people do, or solely as a means of holding together a clique of men who help each other out in business deals. Both of these elements may be present in Freemasonry. But in themselves they do not explain why millions of men in Europe and the United States join Masonic lodges and devote a good deal of time, energy and money to Masonic activities. Groups of this kind are clearly meeting some deeply felt human need, and playing a role of some kind in the wider society that harbors them." [2]

Therein lies the rub. Although Freemasons in the 20th Century knew that the goals and objectives of their Craft were totally harmonious with existing political structures, the veil of secrecy that surrounded the Institution promoted doubt in the minds of the uninitiated. Questions arose concerning why individuals chose to join the Craft and, at first blush, no answers were readily apparent. As Thomas Fuller so succinctly stated: "Would you know secrets? Look for them in grief and pleasure." The anti-Mason saw the Craft as a secret society with its esoteric knowledge harboring some undisclosed plot certain to bring grief to their particular interests. The Freemason, however, viewed the Craft from the vantage of insight and experience, and observed no such devious threat or clandestine purpose. This difference, however, spawned much dissension and strife between various anti-Masonic bodies and representatives of the Craft.

This bifurcation of thought concerning the secrecy in Masonry is not totally unfounded, but reflects a lack of knowledge about the Craft. There are four basic degrees of secrecy in all organizations, but for this treatise's purposes, only two need be examined. A private organization establishes rules and regulations and it restricts its membership. Its internal business transactions are private, and *some* of its activities *may* be kept secret. A secret organization, however, not only limits it membership, it also places extremely strict limitations on all activities in order to remain out of the public view.

Freemasonry has always been a private organization. Membership rosters are kept at every lodge, public Masonic funerals are commonplace and lodges are listed in public directories. It is not difficult for a non-Mason to ascertain who are the Freemasons in his community. Only when outwardly persecuted by totalitarian regimes did Freemasonry wrap itself in secrecy in order to preserve its membership and, thereby, its very existence.

It is easy to forget, however, that what is true and common in the 20th Century was not the *affaire du jour* of the 18th Century. In an era when the Inquisition sent men to their death for espousing free speech and when opposition to the church meant swift retribution, the idea of a fraternity where men of different religions, creeds and beliefs could gather in communication without plotting for or against some man-made structure was literally unthinkable. It is against this background that anti-Masonic thought must be placed when assessing its roots and rationale.

This treatise is not intended to defend Freemasonry from the organized and unorganizaed attacks that have occured during the past half-millennium or act as an apologia for such actions. Rather, the first section shall examine the more salient causes of such sentiments and attempt to ascertain their *raison d'entre.* There appear to be two main streams of generalized anti-Masonic thought in recent history as well as localized incidents that did not have lasting international effect. The most obvious schism that existed occurred between the 18th Century Masonic lodges and the Roman Catholic Church, a rupture that reached its zenith in the 1730's with the first in a series of Papal Bulls and excommunications of Freemasons. A second reoccurring anti-Masonic theme focuses directly on the facts that the Craft is a society of secrets. Historically, this has provided much ammunition of the uninitiated, the rumor-mongers and those individuals who realized that a particular advantage could be gained by publicly opposing the Craft. The rapid proliferation of various psuedo-Masonic organizations during this era confused the public and served to discredit Freemasonry by wrongful association. From the vantage point of the 20th Century it is easy to affirm Aesop's classic: "The strong and the weak cannot keep company," but in the 1700's the distinction had not yet been made as to who was strong and who was weak!

Perhaps the greatest strength of Freemasonry focuses on its inherent characteristic as an educational vehicle. A recent edition of a national newspaper featured a front page article heralding the rise of adult students in the United States' colleges and universities. The article was not an isolated incident; the appearance of adult-oriented programs of instruction are rapidly proliferating as the nation becomes increasingly concerned with focusing educational efforts on postsecondary populations. Indeed, a new term "andragogy" has been coined which concerns itself with the adult learner, as opposed to pedagogy, or the practice of teaching youth.

For Freemasonry, this focus is nothing new. The cornerstone of our Institution is predicated upon the fact that Freemasonry is a course of moral instruction and that instruction is directed not only at newly initiated brethren, but also serves as a form of continuing education for all Masons. In actuality, if we examine the conditions of learning as cited by leading American adult educators, we find that the Masonic form of education has long held the recently identified principles of adult learning: (1) the learner "feeling" the need to learn (2) the creation of a learning climate of physical comfort, mutual trust, freedom of expression and the acceptance of difference. These are the very cornerstone of the Fraternity. Whether the lodge is examined at refreshment, the degree work or when the precepts are taught, Freemasonry applies all of the aforementioned concepts in full measure. (3) The learners perceive the goals of the learning situation to be their goals. By the fact that the candidate sought admission indicates such a persuasion, by their personal oaths at later times reaffirms this concept. (4) The learner participates actively in the learning process. No elaboration is needed on this point. (5) The learning process is related to, and makes use of, experiences of the learner. If point 2 can be labeled the cornerstone of the Fraternity, point 5 must certainly be the ornate portal. Only each individual can ascertain with any certainty whether the lessons of Masonry are carried forward into everyday life. Finally, (6) the learners have a sense of progress toward their goals. This can be documented in two ways: through the internal commitment the candidate makes within himself and, of course, through an outward sign of progression through the degrees. It is the former commitment, not the latter, that should concern all Freemasons.

The key concept, of course, focuses on interaction. Interaction is the key principle involved in any meaningful learning and it rests with the application of the knowledge being transmitted. If the knowledge has application in the learner's life, transfer will occur and the desired outcome may result. Freemasonry deals with those moral issues that confront Masons daily. The immediacy of transfer is readily apparent for those who are serious and willing to assume the responsibility of ready comprehension and lasting impression.

Students of psychology inform us that the mark of a mature person is to be able to defer his gratification. The Masonic lesson for the adult learner stands apparent. Simply witnessing the ceremonies and memorizing paragraphs is not sufficient to achieve the enlightenment available to the studious Freemason anymore than simply reading the Constitution makes an individual a good citizen. Only through living the precepts and concepts of morality will the mature person realize the full fruits of Masonry. "The jurisprudence, learning and moralty of Masonry and not merely the work and ceremonies should have our most profound thought and attention." Like other forms of education, the acquired knowledge must be utilized before its value becomes apparent. The very form of Masonic instruction, allegory and symbolism, lends itself to introspection, thoughtful consideration and analytical processes. The casual or curious candidate who seeks immediate gratification through a cursory examination of Masonic materials and ceremonies will be doomed to disappointment. Like all other endeavors, those that are the hardest to obtain will hold the greatest value. It is not by accident that Freemasons are regularly called to "labor."

A Freemason once wrote the following words in the Scottish Rite magazine, *The New Age.*

> "The lessons men learn in Masonry are interpreted by them as individuals so that they can be applied to their daily life. Each time I view a degree, I get a fresh idea of its meaning. I get so many ideas that they sometimes run together and I find myself asking questions."

The writer of those words, of course, has discoverd the key to true learning. Not to simply accept the new because of a directive, but to seek out, ask...and then enter. The twin psychological factors of motivation and readiness to learn and the spiritual hunger all come together to form the Masonic learner. "Ask, and it shall be given you; seek, and ye shall find; knock, and it will be opened unto you."

The lessons of Masonry are available for the enlightened.

II

Why Join: Some Ancient Reasons

"Tell me thy company and I will tell thee what thou art."

CERVANTES

"Everyman is like the company he is wont to keep."

EURIPIDES

In order to ascertain the reasons why individuals join Freemasonry in the late 20th Century, it is enlightening and perhaps necessary to examine the historical reasons for the development and enlargement of the Craft.

Originally, all masons were operative; that is, the mason was an artificer who actually worked in the construction of buildings, most notably the awesome cathedrals that still grace the continent of Europe. The exact ancestory of the medieval builders will never be traced with exactness, but some degree of credibility must be extended to the lineage that extends from the ancient Roman College of Architects. The tie to the older Dionysian Artificers is less clear and one can almost detect a desire on the part of some Masonic writers to lay claim to a more ancient heritage than perhaps exists.

Nonetheless, the masons who constructed the great edifices of the medieval epoch became known as "Freemasons" because they were "free" from the normal restraints common to most workmen of that era. Whereas Guild or local masons paid taxes, they were not allowed to travel from one location to another without permission and they were almost always in a position of total servitude to local feudal barons. Free-masons were unrestricted by the aforementioned restraints and formed, in essence, a privileged class that enjoyed civil liberties unknown to members of the laboring class.

It was not unusual, therefore, that these free-masons formed associations to not only provide mutual aid and assistance to one another, but also to keep their knowledges and information secret to themselves and those they decided should be allowed to acquire said knowledge. In this manner they could regulate the number of free-masons and thereby ensure their privileged place in society. The protection of secret or esoteric knowledge was not a new idea conceived by free-masons, but rather common practice amongst groups that desired to protect their self-interests. This practice should not be viewed with exceptional criticism because the association permitted a free exchange of ideas and information that might have proved impossible under different circumstances, especially in light of the politically volatile medieval situation. This was, of course, a period when education was difficult to acquire unless one was associated with the church and, at this point in time, the church was extremely parochial about who was able to acquire "its" knowledge.

Although the exact date in history is lost, Freemasonry experienced a profound and fundamental change approximately 700 to 1000 years ago. The use of the term "accepted" mason appears in the Halliwell Manuscript, the oldest written Masonic Constitution yet discovered, written in or about 1390. The Halliwell document alludes to England's Bonnie Prince Edward who, according to the manuscript, "of speculatyfe he was a master." Edward lived in the 10th Century so it appears certain that Speculative Masonry existed at least by that early date. Speculative Masonry refers to members of the Craft who did not operate as working masons as a means of earning their livelihood. Speculative members were "accepted" into the Craft and when combined in the same association or lodges as the "free operative masons," the Craft's transformation was complete. The Craft then became a body of "Free and Accepted Masons" and within one-half millennium that accepted faction far out-numbered their operative counterparts.

The immediate question focuses on why free masons would want non-operative men in their lodges and, conversely, why would non-masons want to join an association of laborers.

There is no simple answer to these poignant queries. Instead, various historians have isolated six basic categories of reasons why individuals, non-operative masons, have traditionally joined the Craft since Speculative Masonry was created.

Before discussing the reasons why non-operative masons desired admittance to the Craft, the question arises why operative masons would want to admit non-operative masons. There were two basic considerations: financial and social. The lodges used non-operative candidates as financial angels or patrons who contributed handsomely to their respective lodges and their coffers swelled accordingly. Perhaps of secondary importance, but a factor nonetheless, was the age-old desire of the working class to gain an equal status with individuals of the so-called "upper" or "leisure" class. Operative masons were laborers who put in the proverbial "honest day's work for an honest day's wage." Perhaps as a corollary to the social consideration was the potential political repercussion if the lodges did not open their doors to the non-operative aspirants. The reason the free masons were "free" focused on the local authorities granting permission for them to travel freely, avoid payment of taxes and other appealing features. If the landed gentry turned against them, chaos would follow. So, a combination of financial, social and undoubtably political considerations entered into the Craft's decision to admit non-operative masons.

The reasons why non-operative masons wanted to join with members of the Craft are more readily apparent and they fall into six basic categories as identified by most historians.

Without a doubt a prime factor of seeking admittance focuses on the belief that the free masons possessed esoteric knowledge of the ancient mysteries. The famous Welsh poet Goronwy Owen and the equally famous antiquarian Dr. William Stukeley openly admitted they sought admittance because they thought they might obtain some hidden wisdom. Elias Ashmole became a speculative mason in 1646 supposedly because he believed masons had esoteric knowledge in alchemy and the Cabala. Sir Christopher Wren was among the first of many architects who joined the Craft in order to plumb the Masonic history for related information. The earliest aspirants, the wealthy nobles, had inherited wealth and position. There was no corollation between their status and their level of education and unless they chose to associate closely with the church their education was undoubtably limited. The opportunity to acquire knowledge, and not simply esoteric knowledge, was obviously attractive and alluring. In later centuries, educated men such as Voltaire, Rousseau, Alexander Hamilton, Louis Blanc and Pierre Joseph Proudhon joined the Craft as the Lodge presented the opportunity to *exchange* knowledge freely without political or ecclesiastic repercussions. It stands *primae facie* that knowledge was a prime factor in attracting non-operative Masons to the Craft; at first simply the opportunity to acquire it and later the opportunity to exchange it. Why the stress on knowledge? Perhaps Daniel Webster most aptly answered this question in an address given after, most appropriately, laying a corner stone in Boston in 1825 when he stated:

> "Knowledge, in truth, is the great sun in the firmament. Life and power are scattered with all its beams."

A second consideration of early aspirants and certainly a prime factor by the 18th Century was the social aspect of Masonry. In eras devoid of mass media, i.e. radio, television, movies, etc., society tended to form social diversions through voluntary associations. At one time Masonic banquets, parades and public installation of officers attracted great followings and much publicity. Jim Roberts chronicles Masonic activities in England during the first quarter of the 18th Century:

> "Beginning with the Grand Feast in 1721, a practice of masonic processions to St. John's Day Banquets had also grown up. These became spectacles of some scale and even magnificence, many coaches taking part: the procession attending the installation of the Duke of Norfolk as Grand Master in 1730 seems to have been especially sumptous. Masons also courted publicity by making official and corporate visits to theatres, at which special masonic material would be inserted...many of these additions and insertions were subsequently given wider circulation when they appeared in journals, songbooks, or as broadsides."[3]

Certainly the pomp and circumstance surrounding the Craft proved magnetic to non-operative Masons. One might draw a parallel between collegians pledging a fraternity or junior executives joining the service club or country club that numbers their senior executives amongst their membership. The socialization aspect was real and cannot be ignored. The fact that this reason for joining does not rest in concert with stated Masonic reasons for seeking admission is lamentable, but does not alter fact.

The third and fourth reasons can be said to develop from the aforestated desire to acquire esoteric knowledge. In addition to simply acquiring the information there was a definite bifurcation. The first faction felt that some material gain could be derived from an association with the Craft. This might come from using the esoteria or it might come from personal contacts made within the

Craft. In either regard, membership in the Craft was predicated not on spiritual or moral reasons, but on anticipation of material gains. The second faction felt that some form of self-development might occur from membership within the Craft. This is not in disconcert with established goals and objectives of the Craft.

A fifth consideration centered on the use of Masonic charity. In an age when the extended family was the prime source of social welfare if sickness or old age impaired one's ability to provide for oneself or nuclear family, the idea of a private welfare system was attractive. The concept of Masonic charity probably grew in direct proportion to the development of Free and Accepted Masonry as a social institution. The final factor appealed to a small populace and it developed around the entertainment value of the Masonic rituals. Again, in an era when all entertainment was generally self-generated, the ability to observe dramas and even participate appealed to a certain segment of the population. As most contemporary Masons can attest, virtually all Masonic rituals, when properly clothed and acted, are the equal to many professional productions.

In capsulation, Masonry as a speculative Craft developed in an age that was searching for the true nature of man and the universe. Many non-operative Masons utilized the Masonic Craft for intellectual and/or moral enlightenment. These aims were compatible with Masonic goals and objectives. Unfortunately, some individuals had more crass objectives and utilized the Craft for more material or worldly gains. Whatever the reasons for soliciting membership in the Craft, the Craft's total membership began to increase geometrically.

III

Why Join: Contemporary Reasons

Undoubtably, there are as many different reasons for individuals joining the institution of Free and Accepted Masons today as there are individuals. Perhaps the best manner in which one can explore the reasons for contemporary man associating himself with the Craft is by first placing the traditional reasons for seeking membership in juxtaposition with contemporary standards.

First and foremost, few individuals, if any, seek membership in order to acquire esoteric knowledge supposedly handed down from generation to generation explaining the ancient mysteries. There are thousands of books and publications available to even the most casual scholar that lay bare all secrets and rituals of Freemasonry. The secrets that remain refer only to certain grips, signs and words; but allusions to mysteries, alchemy and higher knowledges have dissipated. It would be unusual indeed to find a reasonable, intelligent man who believes he can find hitherto hidden cognitive knowledge within the lodge. The knowledge of Freemasonry today rests with moral and spiritual matters; matters that a diligent individual can find within the lodge or elsewhere.

The socialization factor is a quasi-historical anachronism. The historical setting of early Speculative Masonry rested, logically, in a different social climate. Today, men can socialize freely and openly in a variety of settings without fear of political repercussions. More importantly, the social prestige that once attracted membership has subsided. Professional organizations attract philosophers, civic clubs attract social activists and political parties provide a vent for social reformers. Quasi-fraternal organizations such as the Elks, Moose or Eagles provide the opportunity for social gatherings with alcoholic beverages. In capsulation, the social aspects of Masonry now revolve diestic fellowship, morality and altruistic values. Admirable traits all, but in a world that appears to be running amuck with nihilistic concepts, the dearth of members might be anticipated. One could argue, however, that whereas the quantity has been eroded, the quality has raised. That is true, no doubt, but a society that will solicit membership and imposes rather demanding proficiencies before full membership can be attained, will find itself constantly faced with a dwindling populace.

The other historical reasons for membership can be dealt with in a rather summary manner. The material benefits of Masonry still exist, but it is totally contrary to established Masonic regulations to join the Craft for such a crass reason. It happens, of course, but such members usually are noticeable by their absence at regularly scheduled functions and quickly identified as being rather mercenary. The ensuing fraternal ostracism is effective. More to the point: social welfare systems throughout the world have supplanted the obvious need for charity. Although contemporary charity is welcomed and needed, if it failed to exist it is reasonable to assume society would be able to fill that void. The *raison d'etre* of Masonic charity has been paralleled and then surpassed by social security, medicare/medicaid, and other forms of government intervention. They are not as humane, perhaps, but fiscally more encompassing. It is almost an understatement that would be more stimulating to the one who once turned to Masonry as a source of entertainment.

In many regards, what has occurred has been the alteration of the reasons that attracted individuals to Masonry three centuries ago. The higher altruistic and spritual reasons still exist.

It has been a generalized opinion that Freemasonry today appears to be more of an unofficial "guild" of business men, laborers and community leaders—if only because of anticipated "square dealing" amongst its members. This position undoubtedly has some merit as evidence by the large number of individuals who pay their annual dues, but for a variety of reasons seldom attend lodge. To a casual observer, Freemasonry is an efficacious mutual aid society whose relevance to modern life is like that of the rituals of a modern court, a college graduation or any other public ceremony that has its roots in antiquity. But, is this relevant enough?

This writer believes that men choose to join Freemasonry today for a variety of reasons; not all of which are in concert with what the philosophers of Freemasonry consider acceptable.

There is strong evidence that joining any association fulfills what some psychologists call "satisfying motivational needs." Without delving into the depths of psycho-analysis, it can be alleged that there is a hierachy of needs in every man that must be satisfied in order for that individual to enjoy good physical and mental health. Freud, Adler, and Jung attempted to explain man's behavior in terms of single needs, i.e., sex, power or the desire for individuality. Karl Marx felt economic factors were the prime determinant. The school of psychology that speaks to this subject, however, has its roots with men such as Abraham Maslow or David McClelland. The school of thought tends to group human needs and the desire to satisfy those needs in a hierarchy. Once man has satisfied his hunger and thirst, for example, he can seek clothing or a dwelling. Once those needs are met, other motivating factors arise. In a five tiered scheme, lising those needs requiring precedence over the others first, we have (1) physiological needs, (2) safety needs, (3) affection needs, (4) esteem needs, and (5) self-actualization needs.

Whereas operative masons probably fulfilled safety needs first, contemporary speculative Masons do not fall into that category. Most Speculative Masons utilize the Craft and its brotherhood in order to realize greater self-actualization. Because of the moralistic-theistic shadowing of the lodge, it might be safely asserted that this self-actualization occurs in the area of moral and emotional development; that is, an emphasis on learning to cope with higher altruistic values such as truth, loyalty, virtue and fidelity. As Cantrill has put it:

> "In the midst of the probabilities and uncertainties that surround them, people want some anchoring points, some certainties, some faith that will serve either as a beacon light to guide them or as a balm to assuage them during the inevitable frustrations and anxieties that living engenders."[4]

Freemasonry, to the Freemason, offers those anchoring points. Every individual brings to the Craft certain aspirations, wishes, desires and wants. To the extent that Freemasonry can be counted on to provide satisfaction to the individual will directly relate to his continued participation and activity in the Craft.

Freemasonry, as with other reference groups, serves a normative function by enforcing conformity to group standards. To the extent a man identifies and associates himself with Freemasonry, its goals and objectives, the influence of the Craft on the man will wane or increase.

A classic case-in-point focuses on a Freemason I know that seldom missed lodge meetings while living in the same community with his father, who is also a Freemason. When I met this individual he had been living in another community for three years and had never attended a Freemason's meeting once in that time peroid. When asked why, he told me that he had become a Mason because of his father's influence, but did not find lodge relevant to himself anymore. When visiting at home he attended lodge with his father because it made the elder Mason proud. The degree of internalization in this case of Masonic principles is questionable.

This is not an isolated incident. When I took my Blue Lodge degrees there were three other men taking their degrees at approximately the same time. Only one man of those three ever assumed full membership, the other two dropping out before completion. I asked the one individual who finished what was his reason for joining. "Well," he answered, "both my bosses are Masons so I figured it couldn't hurt." I did not have to ask the other two men why they did not finish because I had the opportunity to discuss Masonry with them after lodge one night. Both bemoaned the fact the Masons they had met were elderly, not too friendly and "nothing seemed to happen" at lodge. They did not return. The age factor and Freemasonry is a reoccurring theme in an age of dwindling membership. I attended one lodge where a comment was made that more Masonic work was done at the local funeral home than at the lodge's altar. I visited another lodge where the average age of those members present was approximately 70 years. The youngest member they had was 45 years old.

It is evident that as the general membership in Freemasonry drops, it can be generally catagorized that many individuals are no longer finding progress in self-actualization toward personal growth and development. Since society now provides social security, welfare and civil liberties, many of the original reasons for joining the Craft have dissipated. Other social organizations such as the Elks, Moose or Eagles provide opportunities for socialization, alcoholic consumption and even golf courses. Freemasonry has lost much of its exclusive character that caused the Craft to grow and prosper even one century ago.

So, what causes men to join Freemasonry in the late 20th Century? To many there is a commitment to higher spiritual and ethical values; but to these individuals the Craft keeps its doors closed until initiation. These individuals must ascertain the quality of the Craft from observing Masons at work and at play. It is, to say the least, a difficult benchmark to assess. Others seek admission without really knowing the institution they seek to join. To them, the ceremonies are barren and meaningless. Yet others seek admission for material gain or familial expectations. To those individuals Freemasonry is, as Shakespeare so aptly stated, merely a walking shadow.

IV

Questions of Antiquity and Confusion

"The disadvantage of men not knowing the past is that they do not know the present. History is a high point of vantage from which alone men see the age in which they are living."

CHESTERTON

When did the Institution of Freemasonry begin? No one knows! Various Masonic historians attempt to trace the Craft as far back as the Culdees; others claim its origin with the Rosicrucians; while a third theory places its creation in the secret societies of England that supported the Stuart pretender to the throne. Not satisfied with these theories, attempts have been made to link Freemasonry with the Templars after their suppression in the early 14th Century.

For those not satisfied with these linkages of relative recent origin, some attempt to tie Freemasons to the Roman College of Artificers or even to Encoh, Noah and Adam. Some claims of ancient origin are extremely interesting, e.g. one individual feels he can "prove" Freemasonry originated among the Central American Mayans who, he asserts, transported their rites and ceremonies to the Nile and the Euphrates about 12,000 years ago, and founded the culture of the Middle East.

Not to be outdone, no less an authority than William Preston, a Past Master of London's Lodge of Antiquity, wrote in 1855:

> "From the commencement of the world, we may trace the foundations of Masonry. Ever since symmetry began and harmony displayed her charms, our Order has had a being...no art, no science preceded it." [5]

Even those that seek ancient origin must be satisfied with this landmark. After all, not even Masonry can precede the act of creation!

Freemasonry is not the only fraternity to lay claim to ancient origins. The Modern Woodmen of America, founded in 1883, claim their organization, in various ways, is older than either the Golden Fleece or the Roman Eagle. Norman MacKenzie points out:

> "The Building of King Solomon's Temple plays a central part in Freemasonry...the parable of the Good Samaritan is found...in the Independent Order of Odd Fellows, as is the story of David and Jonathan. The Maccabees related themselves to the Maccabean dynasty of the 2nd and 1st centuries, B.C." [6]

The initiation ceremonies of Freemasonry and other institutions are also designed to emphasize historical antecedents and separate the initiated from the non-initiated. This is usually effected by using a ritualistic drama, with an ancient heritage, in which the candidate is required to answer queries and pass certain types of examinations. Often a symbolic journey is enacted that focuses on the institution's moral teachings. After all the veneer of ritual, ceremony and esoteria have been stripped away, what is stressed is usually inner-illumination of the individual in order to aid in the development of a better man.

Freemasonry, or rather some proponents of Freemasonry, attempt to tie the origins of the Craft to antiquity and, thereby, emphasize the validity of the Institution *vis-a-vis* the so-called wisdom of the ages. In reality, the origins of Freemasonry have gotten inextricably interwoven with the values and mores associated with the Craft. Consequently, some zealots tend to categorize those men in history who exhibit those higher values as being, *ipso facto*, Masons. When queried on the subject, it is usually determined that since these men exhibited such morally desirable characteristics as fidelity, truthfulness, honesty, etc., these men were masons, if not Masons. I have heard one man, at a Scottish Rite Reunion, allude strongly, but stop just short of, stating Jesus was a Mason.

This type of blind allegiance to the past serves Freemasonry no constructive purpose. Indeed, it holds the Craft in an ill light as non-Masons tend to ridicule the idea that Freemasonry claims Adam, Enoch, King Solomon and others as members of their Order. The fact that Freemasonry, as an institution, makes no such claim is lost, as well-meaning Masons assume historically impossible positions.

The simple truth is that Freemasonry's history stands by itself and needs no such embellishment to "glorify" its past. Freemasonry cannot afford to lend credence to the analysis of Jean Cocteau who

THE IRON WORKER AND KING SOLOMON

A Reproduction from the engraving made by John Sartain of Philadelphia and published by Macoy many years ago.

stated, "What is history after all? History is facts which become lies in the end; legends are lies which become history in the end."

If we seek the origins of Freemasonry, a close proximity of that date would be approximately 926 A.D. That is the date cited in the oldest Masonic document yet discovered, the Halliwell Document or *Regis Poem*. But this document alludes only to an organization of builders, not to the specific Craft we recognize as the immediate predecessor of contemporary Freemasonry.

"Freemasonry, as distinguished from any other organization of practical builders probably began among the Cathedral Builders of the Middle Ages, 10th or 11th Century," or so states the Masonic Service Association of the United States. The Middle Ages are far removed from either Enoch or Preston's commencement of the world, but much easier to believe and, more importantly, historically documented.

Of even more recent historical interest is the fact that the first Grand Lodge of Freemasons did not convene until 1717 and the first book focusing on what we know as Freemasonry today was not published until 1723.

Part of the reason for the honest confusion about the origins of Freemasonry stems from the uncritical examination of the Legend of Hiram Abif, the building of King Solomon's Temple and the Legend's origins.

The Legend

If Masonic history dates from the 10th or 11th Century, how does this affect the Legend of Hiram Abif and the construction of King Solomon's Temple? Since King Solomon began construction of the Temple in 967 B.C. and it was finished in 7 years, or approximately 2,000 years *before* the *Regius Poem* places the start of Masonry, the obvious question that arises is whether the Legend of Hiram Abif, as related by Freemasons is a true story. The best answer to that query is found in the Masonic Service Association's publication *One Hundred Questions About Freemasonry:*

> "If by true is meant 'factual' the answer is no. If by 'true' is meant containing a great truth, then the answer is yes...The Legend of Hiram as told in the Master Mason degree is one of the oldest in the world, but Freemasonry's legend is particularly her own. The three who encountered Hiram at the

gates of the Temple are themselves symbols of error, evil and sin; and the story, as a whole, is of the ultimate weakness of such forces against the power of the Great Architect. The word "Abif" is translated both 'his father' and 'my father' with 'father' used in these senses as a patriarch, a teacher, a source of wisdom and not as the actual father of a family."

This is an extremely important fact to recognize. The Legend of Hiram Abif was never intended to be accepted by the Freemason as factually true, but rather is a prime example of the Masonic use of allegory. The confusing aspect of this particular allegory is that it has a factual historical base. There *was* a King Solomon and he did order the construction of a magnificent temple. There were workers at the Temple that could be divided into the categories of apprentice, fellow craft and master masons. To the casual observer, the allegorical segments became lost in the network of historical facts. Most Freemasons will tell you that Freemasonry is a system of morality, veiled in allegory and illustrated by symbols, but many find it difficult to differentiate between allegory and fact.

There are great truths to be found in Freemasonry, truths that are anchored in the great moral teachings passed down through history. These truths have stood the test of time and are no less valid today than they were centuries ago. To claim more than is due is to simply misunderstand Freemasonry and dilute the character of the Mason and the Craft. These truths know no era, but exist as a golden thread through all ages. The perimeters of these truths can be found in Freemasonry's Ancient Landmarks and Freemasonry's Decalogue. The commonly accepted Landmarks, although each Grand Lodge can deviate from this list, focuses on (1) monotheism, or the belief in one God or a singular diety, (2) a belief in immortality and (3) the belief and acceptance of a volume of Sacred Law. The Sacred Law may be the Holy Bible to the Christians, the Pentateuch to the Hebrew or the Koran to the Moslem; the specific faith is totally unimportant. All of the aforementioned three landmarks are tangible requirements of belief or commitment. The fourth landmark is the Legend of the Third Degree. This Legend, however, is to be understood and interpreted in its allegorical entirety. Albert Pike wrote in his monumental work, *Morals and Dogma*, "...Masonry cannot be understood by simply hearing a lecture or reading an essay, but rather requires thoughtful introspection and personal research." The Landmark of Secrecy (6) is discussed at length elsewhere in this work. The 6th and 7th Landmarks are the acceptancy of the symbolism of the operative art and that every Mason must be a freeborn male.

The Decalogue, or Masonic Ten Commandments, are listed simply and clearly by Pike: (7)

1. God is Eternal, Omnipotent, Immutable Wisdom and Supreme Intelligence and Exhaustive Love.
   (a) Thou shalt adore, revere and love him.
   (b) Thou shalt honor Him by practicing the virtues!
2. Thy religion shall be, to do good because it is a pleasure to thee, and not merely because it is a duty.
   (a) That thou mayest become the friend of the wise man, thou shalt obey his precepts.
   (b) Thy soul is immortal! Thou shalt do nothing to degrade it!
3. Thou shalt unceasingly war against vice!
   (a) Thou shalt not do unto others that which thou wouldst not wish them to do unto thee!
   (b) Thou shalt be submissive to thy fortuner, and keep burning the light of wisdom.
4. Thou shalt honor thy parents.
   (a) Thou shalt pay respect and homage to the aged!
   (b) Thou shalt instruct the young.
   (c) Thou shalt protect and defend infancy and innocence.
5. Thou shalt cherish thy wife and thy children.
   (a) Thou shalt love thy country and obey its laws.
6. Thy friend shall be to thee a second self.
   (a) Misfortune shall not estrange thee from him.
   (b) Thou shalt do for his memory whatever thou wouldst do for him, if he were living.
7. Thou shalt avoid and flee from insincere friendships.
   (a) Thou shalt in everything refrain from excess.
   (b) Thou shalt fear to be the cause of a stain on the memory.
8. Thou shalt allow no passions to become thy master!
   (a) Thou shalt make the passions of others profitable lessons to thyself.
   (b) Thou shalt be indulgent to error.
9. Thou shalt hear much: thou shalt speak little: Thou shalt act well.
   (a) Thou shalt forgive injuries.

(b) Thou shalt render good for evil.
(c) Thou shalt not misuse either thy strength or thy superiority.

10. Thou shalt study to know men; that thereby thou mayest learn to know thyself!
    (a) Thou shalt ever seek after virtue!
    (b) Thou shalt be just.
    (c) Thou shalt avoid idleness.

"But the great commandment of Masonry is thus: A new commandment give I unto you: that ye love one another!"

These are the great truths of Masonry; they are timeless, they need no embellishment. The fact that Enoch was not a Mason, the truth that Adam was not a Mason, the knowledge that Masonry did not exist concomitantly with the creation, does not detract from the great truths of Masonry. Indeed, the Legend of Hiram Abif, which causes confusion among many American Masons, might be more valuable if it was understood that not all Freemasons in the world do not accept the Legend literally. The French, for example, at one time understood the Legend to tell the story of the martyr Jacques DeMolay.

The goals and objectives of Freemasonry are not tied to any historical past that stretches belief and credibility. Freemasonry stands by its own merits. It needs no artificial past or psuedo-created history. Examine the Landmarks and the Decalogue. Does the sincere Freemason need or expect more?

V

A Matter of Historical Perspective

Old events have modern meanings;
only that survives
Of past history which finds kindred
in all hearts and lives.
LOWELL, 1868

Nothing appears to incite Freemasons into amicable disagreement more quickly than divergent views on the topic of Masonic history. The fact that Masons are concerned with their Fraternity's historical roots is encouraging because, as Cicero stated in his *Orator*, "To be ignorant of what occurred before you were born is to remain always a child. For what is the worth of human life, unless it is woven into the life of our ancestors by the records of history?"

Many discussions concerning the antiquity of Freemasonry will never be resolved because the individuals involved are basing their arguments on differing views as to what constitutes the origins of the Fraternity. Sovereign Grand Commander Henry Clausen, in the July, 1977 issue of *The New Age,* alludes directly to this bifurcation of thought when he stated, "Our teachings and symbols preceded our formal organization by thousands of years." Too often this subtle difference is overlooked; that is, the distinction between the formal history of the organized Craft and the development of ancient symbols and rituals currently used by Freemasonry is blurred and not separated as some discussers fail to adequately define their terms and, therefore, confuse the issue. Preston's statement is repeated verbatim in Daniel Sickels' 1875 tome *General Ahiman Rezon and Freemason's Guide,*

> "From the commencement of the world, we may trace the foundations of Masonry. Ever since symmetry began and harmony displayed her charms, our Order has had a being...no art, no science preceded it.[8]

The key word, of course, is "foundation." Sickels asserts that the foundation of Masonry was spawned with the creation of symmetry and harmony. This position is defensible according to the approach an individual wishes to assume when defining the perimeters of Masonic history. It is interesting to note, however, that at no point does Sickels, or any other academician, assume or assert that the organized Institution of Freemasonry, replete with rituals, degrees, symbols and dogmas, commenced with the dawn of history. Sickels, and others, speak to the *spirit* of Freemasonry.

When does the history of Freemasonry begin? That depends entirely on how strictly one wants to define the term Freemasonry whether one refers to the development of moral and social virtues, or the codification of rituals, dogmas and constitutions, or even a reference to the nebulous spirit that has been deemed to have passed like a golden thread from Adam to Enoch to Nimrod to present day Masons. The boundaries of this thread, tracing the different theories of evolution from the Culdees, Essenes and Rosicrucians, are interesting, but as each theory is presented and discussed one is constantly reminded of Henry Adam's homily, "History is a tangled skein that one may take up at any point and break when one has unravelled enough."

The simple truth remains that all ancient mysteries possessed degrees or some similar gradations. Most also appear to have secret conventional signals by which members of the different fraternities could recognize one another. It is possible for each individual to unravel that "tangled skein" and take it back to whatever point-in-time is convenient. The results are not conclusive, however, and to paraphrase Puddin' Head Wilson, "the very ink with which history is written is merely fluid prejudice."

Albert G. Stevens, a respected Masonic historian, has enumerated a minimum of nine theories that illustrate this "tangled skein." "Among the many theories as to the origin of modern Freemasonry the following have had many advocates: (1) that which carries it back through the medieval stone masons to the ancient mysteries, or to King Solomon's Temple, (2) not satisfied with the foregoing, that which traces it to Noah, to Enoch and to Adam, (3) the theory that the cradle of Freemasonry is to be found in the Roman Colleges of Artificers of the earliest Christian era; (4) that it was brought into Europe by the returning Crusaders; (5) that it was an emanation from the Templars after the suppression of the order in 1312; (6) that it formed a virtual continuation of the Rosicrucians; (7) that it grew out of the secret creations of the Partisans of the Stuarts in their efforts to regain the throne of England; (8) that it was derived from the Essenes, and (9) from the Culdees. Each individual can untangle the skein to his own satisfaction.[9]

More substantial facts are traceable through relatively recent lodge records, ancient documents and early manuscripts. Primary source documents such as the Halliwell Document or Anderson's *Constitutions* provide the careful historian with more concrete data by which to trace his efforts; but do not answer the queries related to the events which *preceded* the documents themselves!

The pertinent question does not focus on whether the accumulation of data or facts can be established to offset the evidence another individual might marshall forward to support his thesis. As Sov. Gr. Comdr. Clausen stated, "Tracing them (the roots of the Scottish Rite) is a romantic and exciting quest for adventure in the realm of the mind and the spirit." Because of the many points of departure one might take when defining the "start" of Freemasonry, it is impossible to arrive at one rationale universally accepted conclusion. One might similarly argue whether the Declaration of Independence was truly promulgated in July of 1776 or if its spirit, and therefore its very existence, did not permeate free men everywhere commencing with Adam and passing through mankind as a unifying force cementing Aristotle to Voltaire to Rousseau to Locke and finally passing into the hand and pen of Thomas Jefferson.

In essence, amicable debates concerning Masonic history can do nothing but strengthen the Craft as the proponents of various theories are forced to continue their personal research in order to stay abreast of current and historical literature. A study of Masonic history is truly a process of continuing education as the deeper one delves into the past the more apparent it becomes that there is no *one* right answer. There are only a variety of possibilities, each of which has its value based on the appropriate definition of terms. The enlightened Freemason will recognize the richness and difference of his Institution's heritage and not limit himself to one line of inquiry. Instead, he will look deeply into the vastness of the ages and find a universe of explanations as rich and as satisfying as Freemasonry itself.

Whereas it is true that the origins of Freemasonry can cause amicable disagreements amongst members of the Craft, it is equally true that the use of historical terms and phrases in modern lodges can also promote spirited debate and arguments. Once again this *cause celebre* can be used to Freemasonry's advantage. It is, of course, yet another opportunity for continuing Masonic education and to understand Freemasonry through a proper historical perspective. Few Masonic practices are so misunderstood and confused by the general public, and even some members of the Craft themselves, than the use of ritualistic-symbolic titles and phrases in contemporary lodges. The problem focuses not on the titles and phrases per se, but rather on the ill-fated attempts to translate these terms from their historical framework without regard for altered social conditions. As Albert Pike so aptly stated:

> "To understand *literally* the symbols and allegories...is willfully to close our eyes against the Light. To translate the symbols into the trivial and commonplace is the blundering of mediocrity." [10]

Perhaps the classic example of confusion centers on the title "Worshipful Master" used in the Blue Lodge. Even a cursory examination of Masonic literature reveals that this term does not connote any relationship between the individual in question and a supposed metaphysical attitude, but rather speaks to the generic base of the word "worship" which simply means "greatly respected." To the non-Mason, or the uninformed, who does not take the time to understand the history of Freemasonry, or even desire such edification, the titles conjure visions of Amos and Andy, the Mystic Knights of the Sea, or Jackie Gleason wagging his raccoon-tailed lodge hat. They are the source of ignorant jest and misplaced criticism. The critic's frame-of-reference is myopic and reminiscent of the famous dialogue between Sherlock Holmes and Watson. Holmes rather resented the fact that Watson has never read his monograph on the distinguishing of 140 varieties of cigar ash, but when Watson told him that the earth was a globe revolving around the sun, he remarked that that was a singularly irrelevant piece of information. Likewise, the scoffer who attempts to ridicule Freemasonry on the basis of phraseology is singularly myopic and has not come to grips with the substance of the Craft.

As regretable as these incidents may be, they serve as an excellent catalyst for the Freemason to obtain a greater understanding of his Institution. It has been said that a teacher never really understands a subject until he attempts to teach it. The same can be said for the conscientious Mason; he will never fully understand Freemasonry until he studies it and attempts to reach a higher level of self-education. Without attempting to discover the roots of Freemasonry, the Freemason can never begin to dispel the shackles of self-imposed ignorance. Leonardo Di Vinci's truism was never more appropriate: "Nothing can be loved or hated unless it is first known."

Freemasonry, to be lived and loved, must be understood. The traditional analogy relates the Craft not to an essay, to be simply read, but rather to the unwrapping of a problem that requires thought and research. Although the spiritual aspects of the Institution are personal and often defy verbalization, every Mason should make an attempt to explain the salient processes and outward symbols of Freemasonry—if only to himself.

It is not necessary to become a scholar, but to frame some type of explicable reality. Plato separated knowledge into two planes: an upper level of theoretical knowledge that unites itself to permanent ideas, and a lower level of pragmatic knowledge whose *function* is to embody the upper level forms on the level of physical life. Freemasonry links the theoretical concepts of spiritualism to the practical aspects of everyday life. To achieve these aims Freemasonry uses as its vehicle the rituals and ceremonies of its ancient Institution. Attendant with these rituals and ceremonies are titles, phrases and words that have had their original meanings altered in contemporary society and, to the uninformed, appear meaningless. (*e.g.*, "parts and points," "duly and truly," or "hele and conceal.") To mock, however, is a great mistake because as with all great man-made institutions, the intrinsic values of Freemasonry are not analogous to a thin veneer—easily recognized and accepted, but rather like a deep rich grain that must be exposed, inspected and worked before its true beauty and worth become apparent.

It is important to recall that the terminology in question does not stand alone, but is an integral part of the entire Masonic ritual and cannot be lifted from its historical context. The Masonic Service Association cites the thoughts of the Masonic scholar Joseph Fort Newton on this matter:

> "Ritual is the dramatization of belief, hope and spiritual dream. It assists imagination by giving form to what otherwise would remain formless, present-

ing vivid mental images which lend a reality-feeling to what is often abstract and unreal. It is picture philosophy, truth visualized, at once expressing and confirming the faiths and visions of the mind." [11]

To approach a self-satisfying analysis of Freemasonry one must understand the rituals and in order to understand the rituals one must acquire some grasp of Masonic history. Only after this process of continuing education and introspection will the seemingly penetrating questions and comic allusions of the uninitiated prove innocuous. Only then will the ancient rituals and the accompanying phraseology when placed in the proper context provide the means for "ready comprehension and lasting impression." It is at this point that the concepts of Freemasonry are understood, including such issues as secrecy, titling and even allusions to horrendous punitive measures.

VI

Historical Penalties

> *"When a man takes a oath...he is holding his own hands. Like water. And if he opens his fingers then—he need not hope to find himself again."*
>
> ROBERT BOLT

Perhaps nothing has been so damaging, in history, as the misunderstood function of the penalties to be invoked if the oaths of secrecy in the Masonic rituals and ceremonies were violated. The penalties, horrible and gruesome in nature, can only be understood in their historical perspective. However historically valid the penalties might have been to the societies of the Middle Ages, when brought to public scrutiny 400 years later, their mission and interest had altered radically within the Institution of Freemasonry. Unfortunately, the written word offered no explanation as to the historical origins of the sanctions and the non-Mason who learns of some hideous penalty is apt to scoff and ridicule and even some Freemasons who pass through the initiation and were not given sufficient interpretation of the penalties to accurately interpret their purposes are confused and bewildered.

Historically, the Christian faiths at various times have taken a strict view of the inextricable relationship between the physical body and the metaphysical soul. (It was not the Christian faiths alone that held this view, but since Freemasonry was spawned in this European-Christian environment, the specific example is appropriate.)

Various religious faiths of the medieval era believed that a desecrated body, or a body that had been mutilated and, thereby, "incomplete," could not release its soul into a glorious eternal afterlife. It is as if the ancient religions totally acknowledged the validity of Wittgenstein's admonishment in his famous *Philosophical Investigation:* "The human body is the best picture of the human soul."

There were similar religious beliefs about bodies that had expired as the result of violent executions. There were even specific ways in which bodies were to be permanently interred in order to allow the spirit/soul to be released for salvation. For example, a body buried in unconsecrated ground, such as a tidal flat where the tide ebbs and flows twice in 24 hours, would never be able to release its soul to everlasting salvation.

The threat of execution in an era that focused on bodily mutilation was a severe sanction as it alluded to the permanent damnation of the soul. As horrible as the actual practice of "drawing and quartering" was, for example, the theological consequences were much more depressing.

The Freemasons of this peroid were still attempting to maintain control of their esoteric knowledge in order to regulate their Craft and ensure their favorable position in society. One measure they used to attempt to keep this information private was to resort to punishments that "encouraged" the initiated to keep their oaths of secrecy. The threat of a horrible physical death that mutilated the human flesh coupled with potential metaphysical punishments was an effective tool. Undoubtedly, a certain number of Masonic ritualistic executions may have occurred during the Middle Ages. It is equally obvious that as Freemasonry moved from the operative practice into the speculative and as social considerations began to supplant vocational, the historical punishments evolved from actual practice into objects of moral lessons.

There is no evidence that the physical sanctions mentioned in Masonic ceremonies were ever enforced as late as the 18th century. In spite of this lack of evidence, rumors of Masonic inspired tortures or the punishment for revealing ritualistic secrets continued. In America, the two most prominent rumors focused on the disappearance of William Morgan in 1826 and the death of Joseph Smith, the founder of the Mormon Church. Both men were reputed to have violated Masonic oaths by revealing esoteric knowledge. Both incidents are devoid of any incriminating evidence, but the story of Smith, in particular, has achieved notoriety as he reputedly used Masonic ceremonies as a pattern for the Mormon Church's rituals.

Contemporary Freemasons must view their ancient brethren's physical sanctions in their historical setting; but the Craft utilizes different modes of enforcement today. Unfortunately, the sanction of ostracism, reprimand, suspension and expulsion are likely to be ineffective if the individuals involved have reached the point where the penalties are deemed necessary. Social ostracism has been, and remains, the greatest punishment, short of death, that a society can levy upon a transgressor of predetermined norms. Freemasonry utilized this penalty as well as the triune tools of reprimand, suspension (definite and indefinite,) and, as a final step, expulsion. The continued use of the more violent forms of physical sanction in modern Freemasonry ceremonies is just an example of the Masonic use of allegory. The intent of such sanctions is readily apparent if one goes beyond its face value and interprets the message in light of contemporary society.

VII

Landmarks

(Adapted from Mackey's *Encyclopedia*)

Freemasonry has landmarks to guide the Craft's activities. In ancient times it was the custom to mark boundaries of lands by means of stone pillars, the removal of which, by malicious persons would be the occasion of much confusion, men having no other guide than these pillars by which to distinguish the limits of their property. To remove these landmarks was a serious crime since it caused confusion and chaos.

Masons use the concepts of "landmarks" in a different sense, e.g., not to mark physical boundaries, but rather to limit the boundaries of the Craft. The decision to codify landmarks posed and poses a problem since there is a diversity of thought concerning what actually constitutes a landmark. "Dr. Oliver says that,'some restrict them to the O.B. signs, tokens and words. Others include the ceremonies of initiation, passing, and raising; and the form, dimensions, and support; the ornaments, furniture, and jewels of a Lodge, or their characteristic symbols. Some think that the Order has no landmarks beyond its peculiar secrets.' But all of these are loose and unsatisfactory definitions, excluding things that are essential, and admitting others that are unessential."

Mackey, in compiling his list of landmarks, restricted it to those customs that he felt had gained universal acceptance. He further felt that each landmark included had to be of such antiquity that "Both the enactors and the time of the enactment have passed away for the record, and the landmarks are, therefore, of higher antiquity than memory or history can reach. Mackey established two prerequisites of a landmark: (a) its antiquity being an essential element and (b) each landmark is/was unrepealable.

Mackey's 25 landmarks, adapted and abridged are:

(1) The modes of recognition are, of all the landmarks, the most legitimate and unquestioned. They admit of no variation; and

if ever they have suffered alteration or addition, the evil of such a violation of thancient law has always made itself subsequently manifest.

(2) The division of symbolic Masonry into 3 degrees is a landmark that has been better preserved than almost any other....

(3) The legend of the Third Degree is an important landmark, the integrity of which has been well preserved. There is no Rite of Freemasonry, practiced in any country or language, in which the essential elements of this legend are not taught.

(4) The government of the Fraternity by a presiding officer called a Grand Master, who is elected from the body of the Craft, is a fourth landmark of the Order.

(5) The prerogative of the Grand Master to preside over every assembly of the Craft, wheresoever and whensoever, is a fifth landmark....

(6) The prerogative of the Grand Master to grant dispensations for conferring degrees at irregular times, is another and a very important landmark. The statutory law of Masonry requires a month, or other determinate period, to elapse between the presentation of a petition and the election of a candidate. But the Grand Master has the power to set aside or dispense with their probation and to allow a candidate to be initiated at once.

(7) The prerogative of the Grand Master to give dispensations for opening and holding Lodges is another landmark. He may grant, in virtue of this, to a sufficient number of Masons, the privilege of meeting together and conferring degrees. The Lodges thus established are called "Lodges under dispensation."

(8) The prerogative of the Grand Master to make Masons at sight is a landmark....

(9) The necessity for Masons to congregate in Lodges is another landmark. It is not to be understood by this that any ancient landmark has directed that permanent organization of subordinate Lodges which constitutes one of the features of the Masonic system as it now prevails. But the landmarks of the Order always prescribed that Masons should, from time to time, congregate together for the purpose of either Operative or Speculative labor....

(10) The government of the Craft, when so congregated in a Lodge, by a Master and two Wardens, is also a landmark.... The presence of a Master and two Wardens is as essential to the valid organization of a lodge as a Warrant of constitution is at the present day.

(11) The necessity that every Lodge, when congregated, should be duly tiled, is an important landmark of the Institution which is never neglected. The necessity of this law arises from the esoteric character of Masonry. The duty of guarding the door, and keeping off cowans and eavesdroppers, is an ancient one, which therefore constitutes a landmark.

(12) The right of every Mason to be represented in all general meetings of the Craft, and to instruct his representatives... Originally, each Mason represented himself; now he is represented by his officers.

(13) The right of every Mason to appeal from the decision of his brethren, in Lodge convened, to the Grand Lodge or General Assembly of Masons... A few modern Grand Lodges, in adopting a regulation that the decision of subordinate Lodges, in cases of expulsion, cannot be wholly set aside upon an appeal, have violated this unquestioned landmark, as well as the principles of just government.

(14) The right of every Mason to visit and sit in every regular Lodge is an unquestionable landmark of the Order... This right may, of course, be impaired or forfeited on special occasion by various circumstances...

(15) It is a landmark of the Order, that no visitor unknown to the brethren present, or to some one of them as a Mason, can enter a Lodge without first passing an examination according to ancient usage. Of course, if the visitor is known to any brother present to be a Mason in good standing, and if that brother will vouch for his qualifications, the examination may be dispensed with...

(16) No Lodge can interfere in the business of another Lodge, or give degrees to brethren who are members of other Lodges.

(17) It is a landmark that every Freemason is amenable to the laws and regulations of the Masonic jurisdiction in which he resides, and this although he many not be a member of any Lodge. Non-affiliation, which is, in fact, in itself a Masonic offense, does not exempt a Mason from Masonic jurisdiction.

(18) Certain qualifications of candidates for initiation are derived from a landmark of the Order. These qualifications are that he shall be a man-unmutilated, free born, and of mature age.

(19) A belief in the existence of God as the Grand Architect of the Universe, is one of the most important landmarks of the Order... The annals of the Order never yet have furnished or could furnish an instance in which an avowed atheist was ever made a Mason.

(20) Subsidiary to this belief in God, as a landmark of the Order, is the belief in a resurrection to a future life. This landmark is not so positively impressed on the candidate by exact words as the preceding; but the doctrine is taught by very plain implication, and runs through the whole symbolism of the Order.

(21) It is a landmark that a "Book of the Law" shall constitute an indispensable part of the furniture of every Lodge... The "Book of the Law" is that volume which, by the religions of the country, is believed to contain the revealed will of the Grand Architect of the Universe... Masonry does not attempt to interfere with the peculiar religious faith of its disciples, except so far as relates to the belief in the existence of God... The landmark, therefore, requires that a "Book of the Law," a religious code of some kind, purporting to be an examplar of the revealed will of God, shall form an essential part of the furniture of every Lodge.

(22) The equality of all Masons is another landmark of the Order. This equality has no reference to any subversion of those gradations of rank which have been instituted by the usages of society.... but the doctrine of Masonic equality implies that, as children of one great Father, we meet in the Lodge upon the level...

(23) The secrecy of the Institution is another and most important landmark. The form of secrecy is a form inherent in it, existing with it from its very foundation, and secured to it by its ancient landmark. If divested of its secret character, it would lose its identity, and would cease to be Freemasonry.

(24) The foundation of a speculative science upon an operative art, and the symbolic use and explanation of the terms of that art, for the purposes of religious or moral teaching, constitute

another landmark of the Order... All the comparatively modern rites of Masonry, however they may differ in other respects, religiously preserve the [King Solomon's] Temple's history and these operative elements, as the substratum of all their modifications of the Masonic system.

(25) The last and crowning landmark of all is, that these landmarks can never be changed. Nothing can be subtracted from then—nothing can be added to them—not the slightest modification can be made in them. As they were received from our predecessors, we are bound by the most solemn obligations of duty to transmit them to our successors.

It should be noted that forty-nine Grand Lodges of the United States do not accept Mackey's 25 landmarks as the ultimate authority on this topic. Thirteen Grand Lodges: Arizona, Arkansas, Idaho, Indiana, Iowa, Montana, Nebraska, North Carolina, Pennsylvania, Rhode Island, Utah, Virginia and Washington have adopted no list of landmarks. Four states: Alabama, Louisiana, Missouri and Texas accept the "Old Charges" of Anderson's *Constitutions* as their landmarks. Fourteen states: Connecticut, Florida, Kentucky, Massachusetts, Michigan, Minnesota, Mississippi, Nevada, New Hampshire, New Jersey, New York, Tennessee, Vermont, and West Virginia have created and adopted their own lists of landmarks. The remaining Grand Lodges have either formally adopted Mackey's list of landmarks or have adopted them by custom and usage.

Of what importance are landmarks to a Freemason in contemporary society? There can be a twofold response, both of which echo Laotse's *The Character of Tao:* "Stretch a bow to the very full and you will wish you had stopped in time." Masonic landmarks enable the conscientious Freemason to "stop" or "go" in relation to the established dictates of the Craft.

All social organizations need and require boundaries or outward limitations. The social contract theory postulates that at one time man lived wild, free from restraints other than nature's laws. Man was totally dependent upon his own intelligence, skills and expertise. As man evolved, it became convenient for groups to develop, that is, societies developed with individuals living in close juxtaposition. When man began living together the "law of the wild" was gradually replaced by a social conscience. Certain freedoms were abandoned in order for the individual to enjoy communal benefits. Informally, at first, and then formally, rules were established to govern behavior. If the rules were obeyed, society enjoyed a modicum of security. If the rules were ignored, chaos developed until such time a new norm of acceptability was established. Eventually, as societies grew and developed, the rules and laws governing societal behavior grew like a spider's web to encompass virtually all aspects of social life. In discussing the role of law in society, Oliver Wendell Holmes stated, "The law is the witness and external deposit of our moral life. Its history is the history of the moral development of the race."

Thus, the Ancient Landmarks of Freemasonry trace a moral development of the Craft, codifying accepted norms within the Institution. To violate these norms is to call into question the validity of the organization. The principle of the development of laws and rules to correspond with the development of the complexities of society can be Masonically exemplified by examining the "articles" and "Charges" developed by Freemasons to compliment the broad Ancient Landmarks. Robert Gould, in his *History of Freemasonry*, refers to an "Apprentice's Charge" that included: [12]

(1) First, that you shall be true to God and the Holy Church, the Prince, and to his Master and Dame whom he shall serve.
(2) And not to use carding or diceing or any other unlawful gaming or to haunt any Tavern or alehouse there to waste any man's goods without license of his Master...
(3) An he shall not commit adultery or fornication in any man's house where he shall work or be tabled.

The Fellowcraft was admonished:

(1) (to)...love God well, and holy church, and his master and fellows.
(2) (to)...respect the chastity of his Master's wife and his fellows' concubines.
(3) (to)...help one another by instructing those deficient in knowledge and skill.
(4) (to)...obey the (Masons) Assembly on pain of having to forsake the Craft and be imprisoned.

Finally, the Master Mason was enjoined to:

(1) Take apprentices for 7 years, his craft to learn that is profitable.

(2) Shall no thief accept (into the Craft), lest it would turn the craft to shame.

(3) No work he undertake but he *"conne bothe hyt ende and make."*

(4) He ought to be both fair and free and teach by his might.

(5) His apprentices "he him teach" in all the requisite particulars.

It is interesting to note that the aforestated articles and charges are specific to operative proficiencies and moral conduct. Again, as with the Ancient Landmarks, these regulations set definite boundaries on expected behaviors. The complete list of Master Mason's charges focused on teaching subordinates and personal conduct. The Fellowcraft's instructions centered on inter-personal relations, while the Apprentice's charges dealt with basic morality and understanding his subordinate position.

In addition to a physical reason for the existence of Ancient Landmarks and other rules and regulations, the aforementioned codes also provided a psychological benefit. The human mind seeks comfort and security and in *The Education of Henry Adams* it was affirmed that, "Order and reason, beauty and benevolence, are characteristics and conceptions which we find solely associated with the mind of man." Without mental peace, the physical aspects of existence, no matter how fine, can become totally unimportant or even constraining. It is comforting and reassuring to know that the proper behavior patterns will result in expected social reactions.

Dr. Obadiah Harris in a classic philosophical tome entitled *Unitive Spirituality*, speaks to the conquest of individual difficulties. Harris writes: [13]

> "The conquest of individual difficulties is a gift which everyone longs to possess when beset by troubles. The individual casts about in religion, philosophy, psychology and morality for the right answer to this dilemma. He consults sages, ministers, professors and friends. He seeks to extricate himself from present difficulties and forestall future ones and the misery which they cause.
>
> "To surmount difficulties, four things are necessary. First, we need to know what difficulties really represent in the order of existence; second, what are the wrong movements that cause difficulties; third, how to avoid weaknesses

> or giving up in the face of troubles; and fourth, how to overcome difficulties with the aid of spiritual power."

These are the exact points addressed by the construction of the Ancient Landmarks and subsequent codes and regulations. The first three factors are spoken to directly in the laws and the fourth, "how to overcome difficulties with the aid of spiritual power," is the reason for Freemasonry's existence. "A system of moral instruction, veiled in allegories and illustrated by symbols."

It should never be forgotten, however, that Freemasonry is a system of individuality. It is true, Freemasons are bound together for mutual assistance and development, but each individual must find a level of self-identification if the Craft is to be successful. One need only turn to the great philosophical sages of history to find Socrates saying: "Know thyself," Jesus Christ admonishing, "What shall it avail if you gain the whole world and lose your soul?" and the Indian Rishis declaration: "Realize Yourself." The Ancient Landmarks, the various codes, regulations and laws were all developed to enable the individual to achieve the greatest self-actualization through the kind offices of the Institution. It is through Masonic law this self-actualization is to be realized. Moses taught that justice, equality and liberty were the keys to happiness. He told the Hebrews that they should have but one law for themselves and for the stranger without the gate. Masonic laws, likewise, point to the proper direction, but the individual's path is of his own chosing. Landmarks can identify the correct path, but cannot force the traveler to take that route.

VIII

Masonry, Secrecy and Pragmatism

Pragmatism is inherent in Freemasonry. This pragmatism in an era of a speculative Craft is not as readily apparent as might be explained in the treatises of John Stuart Mill or does it lend itself to the public scrutiny of dialectical polemics, but it is pragmatism nonetheless. Paradoxically, the concept of Masonic secrecy shrouds this pragmatic reality like a nefarious mist which is often misunderstood and confused by both the Initiated and the Profane alike.

It is an aged bromide that often seems to serve as an apologia: "Freemasonry is not a secret society, but rather a society of secrets." To the casual observer this is simply a semantic exercise designed to divert potential negative criticism from the Craft. The issue at hand, however, focuses not on potential deception, but rather on time; or rather, the passage of same.

Esoteric knowledge is not parochial to Freemasonry or some medieval epoch, but rather common to many institutions and spans time immemorial. Early Christians protected against cowans and eavesdroppers through practices of concealment. The Apostolic Constitutions, Tertullian's *Apology*, St. Cyril, St. Basil and a veritable plethora more, all cautioned for oaths of secrecy as the prime method of preserving physical safety as well as ensuring an orderly succession of information (Mysteries) from one generation to the next. It was effective, logical and thoroughly pragmatic.

The forces and circumstances, however, that generated the Society of Secrets' oaths and obligations have changed radically in the past half-millennium. It is both logical and reasonable to assume that a vibrant institution that spanned the same time period would also change and develop.

The secrets that caused Clemens to rephrase the proverbial admonition that he who exposed the Mysteries to the general public would also place a sword in the hands of a child, have been preserved for the diligent Masonic scholar in the Freemason's degrees. In the 20th Century, however, the revelation of these secrets would not have a fraction of the impact that would have occurred two millennium earlier. Indeed, the esoteria associated with contemporary Freemasonry is esoteric only because of the self-imposed confines of the lodge room. The content of these mysteries is available in a variety of forms in virtually every library and postsecondary institution in America. The Mysteries are mysteries only in the historical context of their time. It is not the content, but rather the form of the parochial information that underscores the emphasis on secrecy. An obvious question focuses on why should this secrecy be maintained in contemporary society when the historical basis for such practice no longer exists. If the information can be garnered in a variety of ways, why should Freemasonry retain its oaths of secrecy? Isn't this an anachronism? A Tyrannosaurus Rex in a jet age? Where is the pragmatic realtiy in this practice?

Albert Pike alludes to one answer to these queries when he discusses "Tokens" and "Signs" in his tome *Morals and Dogma*. Focusing on the Latin word *tessera*, Pike delves into both its generic definition and then the commonly accepted Masonic usage. *Tessera*, originally defined a square piece of wood or stone used in making tesselated constructs, eventually evolved into a secondary definition that is singularly germane to Freemasonry. *Tessera hospitalis* was a piece of wood cut into two parts and, as a symbol of friendship, when given to two individuals became a pledge of mutual fidelity. To break the *tessera* was to break the bond of fidelity and, therefore, dissolve the friendship.

In contemporary Freemasonry, the oaths and obligations may be regarded as modern *tesserae*. Whereas the inherent esoteric nature of the Craft has long been dissipated by non-Masonic publications, such as *Masonry Dissected*, the bonds of secrecy now allude not to esoteria, but rather to moral values. To break one's Masonic oaths is to break one's *tessera hospitalis* with every other member of the Craft, known or unknown. This does not negate the impact or the importance of Masonic obligations, but simply aids in the clarification of the continued use of ancient practices. This is not new to Masonry, for in the First Degree we are exposed to the principle of explaining our activities so they do not appear barren and meaningless.

As Freemasonry evolved from the operative to the speculative, many adjustments were effected in making the ancient ceremonies more meaningful to contemporary societies. Fidelity, a timeless virtue, is not affected by the passing of empires, changes in government or the rise and fall of ideologies. The *tessera hospitalis* is still meaningful: only the cause and effect have been altered. Pragmatically, the results are identical. It is not what is kept secret, but rather the process itself. As Aeschylus stated in his *Fragments* four centuries before Christ: "It is not the oath that makes us believe the man, but the man the oath."

IX

The Initiation Process: Masons and the Mysteries

"The births of all things are real and tender, and, therefore, we should have our eyes intent on beginnings."
MONTAIGNE, 1580

When Albert G. Mackey defined "initiation" in his celebrated *Encyclopedia of Freemasonry*, he could only vaguely allude to the full historical significance of that term. Mackey stated that "initiation" to a Mason signifies a reception into the Order and traces the term's derivation from the King of Phrygis, the Romans and the Greeks. While interesting and informative, the subject of initiation in general is an historically complex topic, but when understood enables the Freemason to not only come to grips with his Institution's past, but also its contemporary nature.

Initiation, the process, in and of itself, is a segregating concept. Societies that attempt to keep certain knowledges or practices esoteric, or secret, limit their membership to those who submit to some form of initiation. Norman MacKenzie in *Secret Societies* records that: "The Baukaua men mystify and terrorise the female population through a boy's long initiation festival, which includes circumcision. A demon, the women learn, craves their sons' flesh, but by bringing plenty of food they can appease the monster, who will regurgitate the youths he has swallowed. The hoax is bolstered up in various ways...The tyros themselves undergo a severe test. Their seniors lead them blindfolded, to their several months' retreat, where concealed guards frighten them with sounds and weapons... When the novices finally discover the truth, they are forbidden—on pain of horrible punishments—to breathe a word to the uninitiated." [14] MacKenzie continues by asserting that many of the distinguishing marks of a secret society can be found in the initiation process: the initiation ritual, the ordeal, the oaths, the myth or legend that supports the secrecy and the segregation of men from women.

The History of the Ancient and Honorable Fraternity of Free and Accepted Masons, circa 1890, foreshadowed MacKenzie's general work and focused directly on Freemasonry and its possible antecedents. This work pointed out that in ancient mysteries a candidate for initiation passed through preliminary rites and ceremonies; he was subjected to a symbolic death and was later helped by the society to "rise" to a new existence.

The mysteries, and later Freemasonry, had or have funeral-type initiations. The process of achieving a successful second-life was based on a way of existence aided by allegories, emblems and symbols. *The History* cites many examples, but the following two cases are most illustrative:

In explaining the Mysteries of Mithras:

> "To prepare a candidate for initiation, many lustrations were required, with water, fire and honey. He passed 40 days—some say 80 days—of probation, and ended with a fifty days' fast. These were all endured in the recesses of a cavern, in perpetual silence, secluded from all society, and confined in cold and nakedness, in hunger and stripes, and with cruel tortures. We may be sure that in some instances these were attended with fatal effects. When one died under these cruel inflictions and rigid penances, his body was thrown into a deeper cavern and he was never more heard of. According to a Christian writer, in the fifth century A.D., the Christians of Alexandria, having discovered a cavern that had been consecrated to Mithras, resolved to explore it; when, to their astonishment, the principal thing they found in it was a great quantity of human skulls and other bones of men who had been thus sacrificed.
>
> "Those who survived these severe tests of endurance became eligible to the highest honors and dignities, and received a degree of veneration equal to that which was paid to the supernal deities. The successful probationer was brought forth into the cavern of initiation, where he entered on the point of a sword presented to his naked left breast, by which he was slightly wounded, and then he was virtually prepared for the approaching ceremony. He was crowned with olive branches. The olive, in the Mysteries, commemorative of the olive branch brought by the dove to Noah, was the propitious omen that the patriarch and family would speedily emerge from the gloom of the dark to the light of the day; so to the candidate, that he would be able to exclaim: 'I have escaped from an evil; I have found deliverance.' The priests of Mithras, by a like allusion were called Hierocoraces, or sacred ravens, and the oracular priestesses of Hammon, Peleiades, or Doves; while, in consequence of the close connection of the dove and olive, a particular species of the olive was called Columbas.
>
> "He was annointed with oil of ban, which is the balsam of Benzoin, and clothed with enchanted armor by his guide, who represented Simorgh, a monstrous griffin, whose name indicates that it is of the size of thirty birds, and appears to have been a species of eagle, and said to correspond in some respects with the idea of the phoenix. The candidate was introduced into an

inner chamber, where he was purified with fire and water, and then passed through the seven stages of Initiation, which is represented as a high ladder, with seven steps or gates. From the top of this ladder he beheld a deep and dangerous vault, and a single false step might dash him down to instant destruction, which was an emblem of those infernal regions through which he was about to pass. As he passed through the gloomy cavern he saw the sacred fire, which at intervals would flash into its recesses and illuminate his path, sometimes from beneath his feet, and again, descending from above his head in a broad sheet. Amidst all this, distant yelling of beasts of prey, the roaring of lions, howling of wolves, and barking of dogs, would greet his ears. Then being enveloped in darkness profound, he would not know whither to turn for safety, his attendant would rush him forward, maintaining an unbroken silence, toward the place whence the sounds proceeded, and suddenly a door would be opened and he would find himself in this den of wild beasts lighted only by a single lamp. Being exhorted to have courage by his conductor, he would be immediately attacked by the initiated, who, in the forms of the several animals, and amidst great uproars and howlings, would endeavor to overwhelm him with alarm, and he would seldom escape unhurt, however bravely he might defend himself.

"Hurried from this scene into another cell, he was again shrouded in darkness. Silence profound succeeded, and, with cautious step, he was conducted onward to encounter other dangers. A rumbling noise is heard in a distant cavern, which became louder as he advanced, when the thunder appeared to rend the solid rocks, and the continued flashes of lightning enabled him to observe the flitting shades of an avenging genii, who appeared to threaten with summary destruction those who invaded the privacy of their particular abode. These scenes continued until the strength and endurance of the candidate being nearly exhausted, he was conveyed into another apartment, where a great illumination was suddenly introduced, and his strength permitted to recruit, and melodious music soothed his outraged feelings.

"Resting for a time in this apartment, the elements of those secrets were explained and all of which were more fully developed when his initiation was completed. When sufficiently prepared to proceed, a signal was given by his guide, and three priests immediately appeared; one of them cast a serpent into his bosom, as a symbol of regeneration. A private door being now opened, howlings and lamentations were heard, and he beheld in every revolting form the torments of the damned in hades. He was then conducted through other dark passages, and after successfully passed the labyrinth of six spacious vaults, connected by tortuous galleries, each having a narrow portal, and having been triumphantly borne through all these difficulties and dangers by the exercise of fortitude and perseverance, the doors of the Sacellum, or seventh vault, were thrown open, and the darkness changed to light.

"In conformity with these seven subterranean caverns, the Persians held the doctrine of seven classes of demons. First, Ahriman, the chief; second, the spirits who inhabit the most distant regions of the air; third, those who traverse the dense and stormy regions which are nearest the earth, but still at an immeasureable distance; fourth, the malignant and unclean spirits, who hover over the surface of the earth; fifth, the spirits of the vast deep which they agitate with storms and tempests; sixth, the subterranean demons who dwell in charnel vaults and caverns, termed Chouls, who devour the corrupted tenants of the grave, and excite earthquakes and convulsions in the globe; and seventh, the spirits who hold a solemn reign of darkness in the center of the earth. From this doctrine probably emanated the Mohammedan belief in seven hells, or stages of punishment, in the infernal regions; and seven heavens in the highest of which the Table of Fate is suspended and guarded from demons, lest they should change or corrupt anything thereon." [15]

The Jewish Rabbi also developed an elaborate initiation:

"The progress of the candidate through the seven stages of initiation was in circle, referring to the course of the planets round the sun; or more probably, the apparent motion of the sun himself... [16]

The Ancient Mysteries did not have exclusive claim on the rites of the initiation as later fraternal bodies were to discover.

There are scattered references to the various processes of initiation practiced by the "masons" in the three centuries immediately preceding the convening of the first Grand Lodge in 1717. Robert Gould's *History of Freemasonry* records several representations. One, referring to the Stonemasons of Germany, requires some interpretation since at first blush it appears as if the source being cited is speaking about a ceremony of affiliation. Gould correctly points out that there can be no mistaking the ceremony:

"Fallou presents a graphic description of the ceremony, in which he avers that the candidate was blindfolded, half-unclothed, slipshod, deprived of weapons and metals, a cord about his neck, led three times about the lodge; that he then advanced by three upright steps to the master, undertook an obligation on the Scriptures, square and compasses, was restored to sight, shown the three great lights, invested with a white apron and gloves, etc., etc.," [17]

In discussing the French counterparts of the German Stonemasons, Gould continues:

"Another great point of difference was, the French fraternities practised a veritable initiation—a mystic reception—and treasured venerable legends; whilst the affiliation of the German craftsmen was simply a burlesque ceremony, enriched by a certain amount of symbolism." [18]

As the speculative craft grew, adherence to more standardized rigid initiations dwindled. There was a small protest of the abandonment of traditional standards. In 1708 approximately 50 operative masons seceded from the Lodge of Edinburgh and formed a new lodge, the Lodge of Journeymen, No. 8. For one and one-half centuries they set 10% as the maximum number of members who were not operative masons and all candidates for initiation had to take a virgorous physical examination completely disrobed and then be reclad in long flowing robes.

The mark of the Lodge of Journeymen, No. 8, on Masonic history is but a footnote, but it is interesting to note that Gould records the investigations of a Mr. Springett who believed, as late as 1958, that Operative Stonemasons in England were still quietly working a ritual that has been passed on only to operative masons, mouth-to-mouth, from medieval times.

The initiation practices of the Mysteries appear to have been imitated, in a less stringent manner, by the Freemasons. But were they? As the Masons remained an operative Craft it is logical to assume their esoteric knowledge and practices were more firmly controlled; that is, the initiation process was not, as Shakespeare wrote, "full of sound and fury signifying nothing," but rather pragmatic in nature. As the craft became more speculatively oriented, the initiation ceremonies became less pragmatic from a vocational frame-of-reference and more symbolic in demeanor.

It is an inherent aspect of all initiation practices that a candidate undergo certain tribulations that separates the initiated from the non-initiated or profane. Perhaps the one aspect of Masonic initiation that tied the Craft to the more visible vestiges of the Ancient Mysteries were the drastic penalties threatened if the Society's secrets were revealed. Somehow, the penalties created an atmosphere of mystery about the Fraternity that seemed to recall the Mysteries of centuries past. In truth, the initiation ceremonies of the Freemasons of the 18th Century were but a walking shadow of their operative counterparts. To many, the term initiation acquired almost a social connotation.

The elaborate processes of initiation into the Masonic Fraternity appear to have been mitigated somewhat as the Craft acquired a more speculative nature. Henry Wilson Coil, Sr., gives a good description of 18th Century initiations in his excellent historical tome: *Freemasonry Through Six Centuries:* [19]

> "Initiation into the lodges of Scotland appears to have been a very simple proceeding, and there is nothing to indicate the contrary in England, where it seems most likely that the ceremony consisted largely of reading or reciting to the candidate the Ancient Legends and Charges.
>
> "The pretended exposés of rituals which were published with somewhat disconcerting regularity, commencing in 1723, exhibited for the most part catechistical mixtures of symbolism and nonsense, with considerable variations among the several versions. These appear to have served in part as lectures and in part as means of recognition among Masons. There is no

> evidence of any considerable formality or even attempted elegance or impressiveness of ritual until well toward the middle of the 18th century..."

This appears to be quite a departure from the early mysteries when men might forfeit their lives during initiation ceremonies.

By the 20th Century, Masonic initiations in the United States had reached a degree of commonality that a Freemason from one jurisdiction visiting another jurisdiction might not be able to follow the working ritual verbatim but would have no trouble identifying the vast bulk of the activity.

The question remains, however, about the relationship between the initiaiton ceremonies of the Freemason and the Ancient Mysteries.

Is Freemasonry technically a Mystery in the historical sense of that term? By the strictest definition, the answer must be negative. The secret worship of a god or gods after some initiatory process is generally regarded as an adequate definition. The *Dictionary de la Fables* defines a Mystery as: "Secret ceremonies which were practiced in honor of certain gods and whose secret was known to the initiates alone, who were admitted only after long and painful trials, which it was more than their life was worth to reveal." Mackey further elaborates on this point:

> "These mysteries undoubtedly owed their origins to the desire to establish esoteric philosophy, in which should be withheld from popular approach those sublime truths which it was supposed could only be entrusted to those who had been previously prepared for their reception...
>
> "By this confinement of these doctrines to a system of secret knowledge, guarded by the most rigid rites, could they only expect to preserve them from the superstitions, innovations and corruptions of the world as it then existed." [20]

The dysfunctionality of a relationship between the Mysteries and Freemasonry *vis-a-vis* a direct correlation is clearly demonstrable. First and foremost, whereas the rituals of Masonry retain a quasi-esoteric nature, there is no worship, *per se,* within the lodge. There are, of course, volumes of Sacred Law, be it the Bible, Koran, or Veda, but no secret gods are honored and, language not withstanding, one could hardly argue that Masonic penalties, in the physical sense, are a deterent to breaking one's Masonic vows.

Mackey does, however, trace five theories that relate the Ancient Mysteries to Freemasonry. The fact that resemblances do exist cannot be ignored, but they should be interpreted in the historical context of social evolution, that is, as organizations as well as humans, acquire greater maturity and experiences, their present activities will reflect a knowledge of the past. It is within this framework Mackey's following points must be understood. [21]

(1) "Both the mysteries and Freemasonry are derivations from the same common source and Freemasonry is the patriarchal mode of worship established by God himself.

(2) "There is a second theory which, leaving the origins of the mysteries to be sought in the patriarchal doctrines, where (they have been placed)...(is found)...a connection between them and the Freemasonry commencing at the building of King Solomon's Temple.

(3) "A third theory has been advanced by the Abbe Robin in which he connects Freemasonry indirectly with the mysteries, through the intervention of the Crusaders. In the work already cited, he attempts to deduce, from the ancient initiations, the orders of chivalry, whose branches, he says, produced the Institution of Freemasonry.

(4) "A fourth theory...is that as some of them, (Mysteries) especially those of Mithras, were extended beyond the advent of Christianity, and even to the very commencement of the Middle Ages, they were seized upon by the secret societies of that period as a model for their organization, and that through these latter they are to be traced to Freemasonry.

(5) "But perhaps, after all, the truest theory is that which would discard all successive links in a supposed chain of descent from the mysteries to Freemasonry, and would attribute their close resemblance to a natural coincidence of human thought."

It is this fifth point that bears further clarification.

> "The legend of the third degree and the legends of the Eleusinian, the Cabiric, the Dionysian, the Adonic, and all other mysteries are identical in their object to teach the reality of a future life; and this lesson is taught in all by the use of the same symbolism, and substantially, the same scenic representation. And this is not because the Masonic rites are a lineal succession from the Ancient Mysteries, but because there has been at all times a proneness of the human heart to nourish this belief in a future life, and the proneness of the human mind to clothe this belief in symbolic dress."

In summation, the relationship between the Ancient Mysteries and Freemasonry, like most aspects of the Fraternity, is essentially a question of personal interpretation. The process of initiation, however, does not restrict itself to either the Mysteries or Freemasonry, but rather extends to all organizations seeking to create a commonalty of experiences for new members. The greater the desire to keep certain information within the parochial domain of that society, the more rigid the initiation or screening process. Historically, certain emblems, symbols, and allegories appear and re-appear in the ceremonies of various societies. Mackey believed that "natural conditions" promoted Freemasonry's use of emblems, symbols, etc. that also appeared in the Ancient Mysteries. This author concurs with Mackey, but also hastens to point out that the ancient masons were undoubtedly aware of the Ancient Mysteries and most likely adopted their rituals to Masonic ceremonies as the need required.

In either case, and neither can be proven nor disproven, Freemasonry does not suffer from any of the aforestated interpretations. It is wise to remember when conversations begin centering on ancestral origins that although such discourses are interesting, the words of SA'DI in *Gulistan* ring true: "Pride thyself on what virtue thou hast, and not on thy parentage."

X

Concordant Bodies: Past and Present

The highest degree in Freemasonry is the Third Degree or the level of a Master Mason. Of this there is no equivocation, reservation or argument. Although the first 3 degrees represent the pinnacle of personal Masonic enlightenment, Freemasonry does not restrict itself to the three historically entitled "Blue Lodge" degrees, however, and concordant, or harmonious, bodies co-exist with Blue Lodges. Contemporary American Freemasonry finds two concordant bodies: the York Rite and Scottish Rite assuming dominant roles, but these are not the only associations with Masonic ties and, historically, many bodies have appeared to compete with these two organizations.

Rival Masonic bodies, quasi-Masonic bodies and even anti-Masonic bodies grew and sometimes prospered after the creation of the first Grand Lodge in England in 1717. At this point in Masonic history such terms as profane, irregular, regular or clandestine did not exist and it was possible for Freemasons to move in and out of most competing bodies without fear of Masonic expulsion and social ostracism. Obviously, this situation has altered considerably in the succeeding two and one-half centuries, but it is interesting and instructive to review a few historical antecedents of modern concordant bodies before examining the salient contemporary associations.

Many historians label an early group of Masonic organizations as "Dissenting Masons." The organizations did not attempt to use the existing Blue Lodge structure as formalized by the new Grand Lodge, but often disputed the very organization created by that 1717 body.

H.W. Coil succinctly capsulizes these dissenting organizations: [22]

> "Our knowledge of all these bodies is scant. Most of what we know comes from scattered newspaper notices. The six or seven to which we find such references were: (1) Philo Musical et Architecturae Societas Appolloni; (2) Appollonian Masons; (3) Antediluvian Masons; (4) Honorary Masons; (5) Real Masons; (6) Modern Masons (7) Scald Miserable Masons."

As Coil states, our knowledge of these dissenting Masons is slight, but a further examination of the "Scald Miserable Masons" is interesting. This was a loose organization formed about 1740 and whose historical notoriety stems mainly from its anti-Masonry bias. It was not a body of Masons, but mock-Masons whose claim to (sic) fame focused on a processional that was reported thusly in a March issue of the *London Daily Post:* [23]

> "Yesterday, some mock-Freemasons marched through Pall Mall and the Strand as far as Temple Bar in procession; first went fellows on jackasses, with cow's horns in their hands; then a kettle-drummer on a jackass, having two butter firkins for kettle-drums; then followed two carts drawn by jackasses, having in them stewards with several badges of their order..."

Evidently the mock processional and a group of Freemasons met at the Temple Bar where the *Post* reports "agreeable" humor was exchanged. Exactly how "agreeable" the Freemasons accepted the processional might be discerned from the fact that a few days after the parade, several of the "mock masons" were conscripted into the British Army.

The Scald Miserables are a relatively unimportant footnote in Masonic history, but both their name and the form of their activity are interesting and their very existence speaks to the growing public awareness of Freemasonry.

There is yet another classification of societies that were not Masonic, but appear to have organized, among other reasons, simply to embarrass or supplant existing Masonic orders.

Among the more interesting of these organizations was the "Ancient Noble Order of the Gormogons." The Gormogons claimed to have descended from the First Emperor of China and required Freemasons to renounce Masonry in order to join their order. At one point the Gormogons announced that a "Mandarin" of their order was in Rome and the Pope and several Cardinals "thought well of their Order." With the exception of one notable Freemason, few joined their association and by the mid 18th Century the Gormogons had virtually disappeared. What is particularly interesting are the theories surrounding the creation of the Gormogons. One school of historians believe the Jesuits sponsored the organization in order to combat the growing influence of Freemasons in European societies. A second body of thought places the creation of the Gormogons with the infamous Chevalier Ramsay; also, as an anti-Masonic device.

As with most groups that are formed to oppose another force, most of their energies are dissipated in negative pursuits and positive features are relegated to a secondary nature. Most dissenting or anti-Masonic organizations lived sporatic lives and except for localized cases, i.e., the William Morgan affair, did not greatly influence the course of Freemasonry's development.

Of more lasting impact to Freemasonry was the development of "side" or "honorary" degrees. These degrees are neither sanctioned by a Grand Lodge nor do they form a part of any regular ritual. The principal side degrees, as enumerated by Albert Mackey are: [24]

1. Secret Monitor
2. Knight of the Three Kings
3. Knight of Constantinople
4. Mason's Wife and Daughter
5. Ark and Dove
6. Mediterranean Pass
7. Knight and Heroine of Jericho
8. Good Samaritan
9. Knight of the Mediterranean Pass

The various side degrees have interesting histories and a few are quite ancient in origin. The principal purpose focuses on a form of moral instruction. It is interesting to note that not only are the side degrees designated for Master Masons or other Masons, a few such degrees are reserved for females who are connected, by blood or marriage, to Freemasons.

A good example of a side degree is "The Secret Monitor." This degree's entire focus centers on the practical elucidation of the Masonic virtue of brotherly love. Extensive use is made of the First Book of Samuel.

"The Mason's Wife and Daughter" is an example of a side degree conferred upon mothers, sisters, wives and daughters of Masons which enable them to obtain the aid of the Fraternity. It is sometimes called the "Holy Virgin Degree."

There are various Orders whose members are composed of relatives of Freemasons. Some Orders are androgynous, that is, they are conferred upon both men and women, others are solely for women. A few of these orders are: The Order of the Eastern Star, Order of the Amaranth, Daughters of the Nile, White Shrine of

Jerusalem, and the Ladies Oriental Shrine of North America. For the youth of non-Masonic age there are, among others, Order of De Molay for males, Order of Rainbow for Girls, Order of Job's Daughters and the Daughters of the Eastern Star for females.

Before examining the Scottish Rite and the York Rite, it is necessary to cite other American Masonic bodies and research lodges. Much like numerous "side degrees," many other Rites and Orders of Freemasonry in the United States cannot be petitioned for, but gain their membership by extending invitations. One acquires membership in these groups by effecting research work in Masonry and, thusly, becoming known to the current members. A few of these Orders are: Grand College of Rites, Masonic Order of the Bath, The Blue Friars, and Allied Masonic Degrees. Some of the Orders confer degrees, others control degrees but never confer them and, yet, others are purely honorary.

Perhaps the most popular concordant Masonic bodies in the United States are the Scottish Rite and the York Rite. The use of the word "rite" in Masonry is derived from the Latin *ritus* or "an approved usage or custom, or an external observance." Masonically, a rite is used as a method of conferring Masonic light. Mackey lists 37 different Masonic Rites to many of which the casual student of Freemasonry has never been exposed, namely: (1) the Reformed Rite, (2) Rite of Martinism, (3) the Swedish Rite, (4) the French Rite or (5) Fessler's Rite. It was a combination of circumstances, planning and organization that elevated the Scottish and York Rites to the positions they hold today.

The Scottish Rite was formed in America at Charleston, South Carolina on May 31, 1801. Allusion to a more ancient heritage, namely by Frederick the Great of Germany, have been made but never supported by historical documentation. The earliest recordable date for the Rite's creation was probably 1754 in France. Reputedly, Chevalier de Borneville of the College of Jesuits formed a Chapter with 25 degrees. During its early history there were several rival groups, namely, the Cerneau Rite and the Rite of Memphis, and the Scottish Rite did not assume a dominant position until it came under the mentorship of Albert Pike. Today, there is a Northern and a Southern Jurisdiction for the Scottish Rite and neither confer the first three "Blue Lodge Degrees." With some modification, both jurisdictions offer the 4th through 33rd degree separated into 5 divisions: the Lodge of Perfection, the Chapter of Rose Croix, the Council of Kadosh, the Consistory of Sublime Princes of the Royal Secret and the Supreme Council.

The Southern Jurisdiction of the Scottish Rite has traditionally assumed strong positions *vis-a-vis* public funding not going to non-public schools, the compulsory use of English as the language of instruction in the grammar schools and in the concept of the entire separation of Church and State.

The Northern Jurisdiction, it appears to this writer, assumes a more philosophical posture believing in "principles rather than programs. Principles unite men, programs divide them."

In both cases, however, both Jurisdictions attempt to further the aim of Freemasonry through their own interpretations of Masonic goals and objectives.

The York Rite is generally regarded as the oldest of the Masonic related Rites, but has been altered considerably since its inception. Originally, the York Rite consisted of only 3 degrees: the Entered apprentice, the Fellowcraft and the Master Mason. By the mid-18th Century this Rite, in its generic form, had been corrupted. In America, "higher degrees" had been superimposed on the basic three degrees and the York Rite began assuming the form it has today. Contemporary. York Rite Masonry can be divided into 3 sections, (1) The Chapter, which culminates with the Royal Arch degree, (2) The Council, (3) and the Commandery. The Knights Templar associated with the York Rite have no connection with the medieval Knights Templar although many writers have attempted to claim such a succession.

Another Masonic related organization is the Ancient Arabic Order Nobles Mystic Shrine or, as it is commonly called, "The Shrine." The only route into the Shrine is through completion of the Consistory in the Scottish Rite or the York Rite's Commandery. The Shrine is a more socially oriented organization with a pronounced effort in the area of supporting burn centers and hospitals for children.

No mention of concordant bodies would be complete, even a cursory overview, without further reference to the Order of the Eastern Star. The Order of the Eastern Star was not the first attempt to establish an organization composed of relatives of Freemasons, but rather followed the concept established by such groups as the Lodges of Adoption, True Kindred, Constellations, and Families.

The Eastern Star was the product of Rob Morris of Kentucky who received the degrees of another androgynous organization circa 1850. Morris produced two abortive Orders before creating the Eastern Star; the first was called "Constellations" and the second "Families." "Constellations" proved to be too elaborate for the average layman to dramatize and "Families" did not possess enough substance to hold the interest of its members. Morris then turned over his material to Robert Macoy who worked the degrees into the present Chapter system of the Order of the Eastern Star.

There are many Freemasons who believe the Order of the Eastern Star serves as a catalyst to unite Masonic families who might otherwise not have this unique opportunity for moral instruction due to the nature of Masonry. Indeed, the entire concept of androgynous Masonic organizations presents, in a microcosm, a clear perspective on the ramifications of the change from an operative craft to a speculative fraternity.

XI

Freemasonry and the 18th Century Roman Catholic Church

"The surest sign of the estrangement of the opinions of two persons is when they both say something ironical to each other and neither of them feels the irony."

NIETZSCHE (1878)

The 18th and 19th Centuries were extremely important in changing Europe into a more democratic continent respectful of civil liberties. Various nationalistic secret societies served as agents of political change and it would be ludicrous to assert that all Masonic lodges remained totally aloof from these activities. Indeed, a classic case in point occurred in Italy where the Carbonari was a secret society organized to thwart the aims of Bonapartism. Their function was to stir up opposition to the French influence and to press for constitutional reforms. Many Italian Freemasons were also members of the secret Carbonari. It takes no dialectical genius to follow the Church's line-of-thought: the Bonapartist supported the Catholic Church, and in the 18th Century, the Church for all intents and purposes ruled Italy. The Carbonari, a secret society, opposed the Bonapartists, therefore, *ipso facto*, the Carbonari opposed the Church. Since many Carbonari were also Freemasons, the Freemasons must oppose the Church. The Church, therefore, should oppose Freemasonry. There is, of course, a serious fault in such "deductive" reasoning, but to the uninitiated who were unaware of Freemasonry's objectives, the conclusions appeared both logical and reasonable.

The year 1738 marked the first official anti-Masonic act by the Catholic Church, but perhaps the first recorded expressions of concern over Freemasonry's use of secrecy occurred in 1698 when a circular was distributed in England that admonished "all Godly people" to beware of Freemasonry because "members of the craft were anti-Christ." What rationale prompted this admonishment? The flyer asked why else would Freemasons, "a devilish sect, meet in secret places, use secret signs, and take steps to ensure that none observe them. Are not these the ways of evil-dom?" This criticism appeared localized, however, and did not affect the Craft in a pronounced negative way.

Freemasonry did not attract a great deal of publicity until Dr. Anderson published his now famous *Constitutions* in 1723. In the decade thereafter, the Craft grew more open in publicizing their presence. Commencing with the Grand Feast in 1721, a practice of widely publicized St. John's Days banquets had also been established. Masons, often prominent citizens, made group visits to theaters, made speeches and attended other public functions. References to the Craft began appearing in journals, newspapers and even dramas. In 1731 a comic-farcial opera entitled *The Generous Freemason* was performed in London. Although the theme was not favorable to Freemasonry, the factor of importance centered on the public examination of the Craft.

Whether the public display on the part of the Freemasons prompted the criticism, or whether the criticism prompted the Masons to react publicly will never be known. What did evolve from the early publicity was severe criticism directed toward the Craft that could not be refuted without inviting open inspection of the lodges and ceremonies. Since such open inspection was impossible, the accusations and charges stood virtually unchallenged and the rumors spread unabated.

The attacks were not gentle. "Freemasons: An Hudibrastick Poem," published circa 1723, hinted at Masonic practices of sodomy and flagellation. The charges appeared to have validity since women were excluded from lodges.

In 1726, a pamphlet entitled the *Freemasons Accusation and Defense* told of a father's attempt to dissuade his son from joining the Craft on grounds of alleged practice of sodomy that occured within the lodges. Three years earlier, in 1723, *Ebrietatic Encomium* brought charges of drunken debachery against Freemasons. Thus, the Masons at the turn of the 18th Century found themselves characterized as sodomists, drunks and otherwise quite a lot of despicable characters.

All of these attacks, however, appeared to do little permanent damage and as the Craft attempted to refute them their membership continued to grow. Only one publication of this era, *Masonry Dissected* by Samuel Prichard, stands by prima facie evidence. Prichard presented the first detailed narrative of the rituals, initiations and oaths and the book was reprinted in 30 different languages by 1800. It was, truly, a dissection of Freemasonry that laid bare previous esoteric knowledge. Prichard's pamphlet caused concern because by mentioning the Masonic oaths and the consequences of violating the vows of secrecy, people began to question the "innocence" of Masonic lodges. Why would such terrible retributions be heaped upon individuals who participated in harmless ceremonies. Exposure of the legend of Hiram Abif was interpreted as a Masonic pledge of revenge. Unfortunately, for the Craft, explanations of symbolism and allegory fell upon deaf ears.

The important factor to mention focuses on historical evidence that early criticisms centered not on purely political or anti-church grounds. Secrecy was the key and, although the charges were damning and embarrasing, they proved to be neither fatal nor destructive by themselves. Coupled with political developments one-half century later, they served as a foundation for anti-Masonic persecution. By 1723, a foundation for opposing Freemasonry had been laid. In that year Pope Clement XII published *In Eminenti*, a Papal Bull excommunicating all members of the Craft and reserving to the Pope himself the power to absolve them. In 1751, Pope Benedict XIV reissued Clement's Bull in his own *Providas* and in 1789, Pope Pius VI reissued *In Eminenti.* The Church's position had materialized, but why did it assume this position? Officially, the Church's stance in *In Eminenti* stated that the association in the Craft of different faiths was, in and of itself, bad. Further, it mentioned that "some people of judgement" considered the society to be depraved. Thirdly, the Pope contended that the oath of secrecy threatened the Roman Catholic confessional and undermined the authority of the Church. Finally, there was a reference to "other just and reasonable motives known to us." There was no further explanation.

There are understandable concerns of the Church as the Pope could hardly be expected to welcome an announced intention to assemble persons of different religious faiths. He also could not be asked to tolerate a secrecy that he imagined endangered the practice of confession. It should not remain unnoticed either that by reserving absolution of the excommunication to himself, and not a village priest, the Pope would be able to identify Freemasons who he felt opposed his rule and might otherwise remain unnamed.

What would cause the Catholic Church to reach the point of issuing the Papal Bull? The aforementioned difficulties between the Carbonari and the Church was a prominent factor. Also, independent Italian nobles in the northern provinces were also attempting to exercise a larger degree of political autonomy than the Church, at that time, wanted to permit. It was a widely-held position in Rome, but again unsubstantiated, that the rebellious nobles were hatching their anti-church plots in the secrecy of the Masonic lodges. As preposterous as that sounds to the contemporary Freemason, cognizant of the prohibitions of political and religious discussions in a Masonic lodge, this argument carried much credibility in Rome; probably because there was more than a grain of truth in the accusation.

Throughout the 18th Century, Italy experienced political upheavals and, not coincidentally, there were three major attacks on Italian Freemasons. The aforementioned Papal Bulls all excommunicated members of the Craft and required personal absolution by the Pope. What better way to identify unknown opponents than by making them appear before you asking for forgiveness? In order to fully understand anti-Masonry in Italy in this era, one must study Church and State relations. This writer shall not delve into the intricacies of that study, but in order to ascertain the mood of that period it is only necessary to examine the statement of Cardinal Corsini, a confidant of Pope Clement XII (1730-40) who publicly announced "...that society (Freemasonry) was formed in England as a game of innocent amusement but...it had degenerated in Italy, where it has become a school of ungodliness." Hardly a sterling analysis of the Craft, its purposes and activities. By 1738 the Church supported inquisition alleging a connection between Freemasonry and freethinkers at the University of Pisa; and later that same year a Masonic bookseller in Florence had his establishment raided and prohibited books were destroyed. With the death of the Medici Pope in 1737, the struggle between Rome and Florence for the Papal throne increased the anti-Masonic activities, and Freemasons were persecuted by both factions as supposed protagonists for the opposite party.

In this particular case, the anti-Masonic activities in Italy paralleled similar actions in France. In the 18th Century England, which spawned Freemasonry, had a history of limited tolerance for diverse political and religious views and had no legal tradition of prejudice against voluntary associations. The populaces of France and Italy, however, did not enjoy the same degree of civil liberties at this point-in-time, and individuals suffered for their beliefs if they did not coincide with the governing clique. Freemasonry was regarded with suspicion from its inception in Roman Catholic Italy and France if for no other reason that its genesis occurred in Protestant England. It should be pointed out, however, that even in England, in the late 17th Century, the English government cast a jaundiced eye on Freemasonry because some lodges housed influencial Englishmen sympathetic to Jacobite and Roman Catholic causes. Evidently, the sword had a double cutting edge and the Masonic Order often received both sides.

XII

Masonry and Literature

"The pen is the tongue of the hand—a silent utterer of words for the ages."
HENRY WARD BEECHER (1887)

It should not come as a surprise that a fraternity that included many of the greatest socio-political activists and philosophers of recent history should also find its name, outward customs and public rituals recorded in the pages of classical and popular literature. Freemasons such as Voltaire, Christopher Wren, Edward Gibbon, Elias Ashmole and Roscoe Pound were easily recognizable and their actions, associations and beliefs influenced societies far beyond the confines of their native lands. These men, plus countless others, exemplified the morals and dogmas of Freemasonry and provided the author, poet and antiquarian of all eras the opportunity to use the Craft as a source of creative inspiration, knowledge and historical reference.

Rudyard Kipling is a prime example of a Freemason who incorporated the Craft's teachings and symbols not only into his personal life, but also in his writing. Kipling often expressed the sentiment that Freemasonry not only confirmed his faith in a common fraternity of mankind, but also reinforced his belief in the desirability of pursuing a common moral code. Kipling's works are punctuated by references to the Craft if the reader is alert to the allusions and descriptions contained therein. Two short stories and poems stand most prominent *vis-a-vis* Freemasonry. In the tale, "The Man Who Would Be King," Kipling describes two Master Masons who established a kingdom in Kafiristan by using the rituals and ceremonies of Freemasonry in concert with the native population's "pagan" rituals. The Masonic tone of this work is established at the onset when one Mason asks the other: "Where have you come from?" and the other answered, "From the East and I am hoping that you will give me the message on the Square." Kipling's second story, "In The Interest of the Brethren," is a more heart-warming tale describing the war-ravaged brethren who find and frequent a lodge of instruction enjoying the rituals and fellowship practiced there. As one Mason so aptly stated, at the lodge they experienced: "The Fatherhood of God, and the Brotherhood of Man; and what more in Hell do you want?"

Kipling's poem, "Banquet Night," prefaces the story "In The Interest of the Brethren," both of which are found in the larger book *Debits and Credits*. Its direction can be best described by citing one of the verses: [25]

"Carry this message to Hiram Abif
Excellent Master of forge and mine:
I and the Brethren would like it if
He and the Brethren will come to dine,
(Garments from Bozrah or morning dress)
As Fellow-Craftsmen—no more and no less."

Kipling's, "The Mother Lodge," is my personal favorite. It illustrates the mingling of men of different colors, creeds and social classes on the common level of the lodgeroom floor. Two stanzas still strike a responsive note with contemporary Freemasons: [26]

"Full oft of Gov'ment service
This rovin' foot 'ath pressed
An' bore fraternal greetin's
To the Lodges east an' west
Accordin' as commanded
From Kohat to Singapore,
But I wish that I might see them
In my Mother Lodge once more!

and the refrain:

Outside—Sargeant! Sir! Salute! Salaam!
Inside—"Brother," an' it doesn't do no 'arm.
We met upon the Level an' we parted on the Square
An' I was Junior Deacon in my Mother Lodge out there!

Kipling's autobiography graphically ties Freemasonry to his personal life. [27]

"In '85 I was made a Freemason by dispensation...being under age, because the Lodge hoped for a good secretary. They did not get him, but I applied, and got Father to advise, in decorating the bare walls of the Masonic Hall with hangings after the prescription of Solomon's Temple. Here I met Muslims, Hindus, Sikhs, members of the Arya and Brahmo Samaj, and a Jew Tyler, who was priest and butcher to his little community in the city. So yet another world opened to me which I needed."

Obviously, Rudyard Kipling's Freemasonry was inextricably interwoven with his professional career.

Other noted writers used the subject of Freemasonry as a reference to develop or elaborate a theme. Lew Wallace, the author of *Ben-Hur*, developed the scenario first alluding to the antiquity of the Master in another novel *The Prince of India*, by having the protagonist excavate the hitherto undiscovered tomb of Hiram, King of Tyre. Inside the tomb was found King Solomon's sword and "the instruments sacred then and ever since to Master Masons: a square, a gavel, a plummet, and an inscribing compass." It is interesting to note that the author placed the working tools of a Mason with King Hiram's other treasures: namely, gold, silver, and jewels.

Edgar Allen Poe included a passing reference to Freemasonry in a brief passage from *The Cask of Amontillado*. [28]

> "I broke and reached him a flagon of De Grave. He emptied it at a breath. His eyes flashed with a fierce light. He laughed and threw the bottle upwards with a gesticulation I did not understand. I looked at him in surprise. He repeated the movement—a grotesque one.
> "You do not comprehend?" he said.
> "Not I," I replied.
> "Then you are not of the brotherhood."
> "How?"
> "You are not of the Masons."
> "Yes,yes" I said, "yes, yes."
> "You? Impossible! A Mason?"
> "A Mason," I replied.
> "A sign," he said, "a sign."
> "It is this," I answered, producing from beneath the folds of my *Roquelaire* a trowel.
> "You jest," he exclaimed.

One can only imagine what type of Masonic sign Poe envisioned that would be both grotesque and identifying. Poe was not the only scribe that described the symbols and rituals of Masonry in quasi-unflattering terms. The immortal Charles Dickens satirized Freemasonry in *Barnaby Rudge* and Mozart's "Magic Flute" attempts to link Freemasonry in rather uncomplimentary terms to certain ancient Egyptian traditions.

Even Sherlock Holmes was cognizant of Freemasonry. In a light scene that speaks more of Masonry's stature in public consciousness, A. Conan Doyle presented the following scene in "The Red Headed League." [29]

"Sherlock Holmes' quick eye...noticed my questioning glances. Beyond the obvious facts that he has at sometime done manual labor, that he takes snuff, that he is a Freemason, that he has been in China...

Mr. Jabez Wilson startled up in his chair, with his forefinger upon the paper, but his eyes upon my companion.

"How, in the name of good-fortune, did you know all that, Mr. Holmes?" he asked. "How did you know, for example, that I did manual labor. It's as true as gospel, for I began as a ship's carpenter."

"Your hands, my dear sir. Your right hand is quite a size larger than your left. You have worked with it, and the muscles are more developed."

"Well, the snuff then, and the Freemasonry?"

"I won't insult your intelligence by letting you know I read that, especially as, rather against strict rules of your order, you use an arc-and-compass breastpin."

"Ah, of course, I forgot that."

A minor passage to be sure, but indicative of the notoriety Masonry enjoyed during this era. A more penetrating example can be found in Count Tolstoi's epic *War and Peace.* A moral thread of redemption of the novel's main character is woven through the book starting with his initiation into a Russian lodge. Robert Burns, himself a Freemason, wrote a stirring ode, "Farewell to Tarbolton Lodge." Goethe wrote Masonic songs, a poem entitled "A Mason's Ways" and a novel that many believe to be the finest Masonic tome ever written: *Wilhelm Meister.* Another German, Gottfried Lessing, utilized the Craft in his novel *Nathan the Wise.* Parallel association can be related to Sir Francis Bacon's *House of Solomon.* The list of Masonic-related literature is obviously enormous and one can only speak of the most salient examples.

The important concept in tracing Freemasonry through literature focuses not on what was written, but rather the fact that the Craft was a focal point of discussion. Freemasonry then, and now, conveyed a message even to the uninitiated and the uninformed. Tolstoi used it to show morally desirable behavior as opposed to moral decadence; Kipling saw it as a vehicle for leveling artificial social standards and Wallace used the Craft for conveying quasi-religious messages. Burns wrote of freedom, harmony and the Grand Architect coming to life in and out of the lodge. Others, perhaps, were more mundane in their usage appealing to outward signs and symbols.

In addition to the inclusion of Masonic signs and phrases in general literature, there are countless volumes of Masonic histories, guides, ciphers and books which focus on each author's personal line of expression. Many of these publications lost their objectivity as the authors endeavored to bring forward a particular sentiment. Indeed, even Dr. Anderson's famous *Constitutions* appear somewhat skewed under the scrutinizing eye of history; for example, the language used could easily be interpreted in such a manner that promised more than was forthcoming. As J. M. Roberts stated: [30]

> "...it encouraged both suspicion among the profane and over-optimism about future revelations among the credulous initiated. Such language thus contributed to a general psychosis of secrecy which gripped those who distrusted the craft...Finally, Anderson's language also favoured the elaboration of systems of 'higher degrees' whose existence was suspected perhaps even before they were devised."

Close on the heels of the publication of Anderson's *Constitutions* came Samuel Prichard's *Masonry Dissected* and, of course, a later book entitled *The Defense of Masonry* by Martin Clare. Two decades later Pope Clement XII's bull of excommunication, *En Eminenti* caused even more agitation and, eventually, pro and con Masonic publications. The plethora of literature thus commenced.

In 1772, William Preston published what is generally regarded as the first Masonic monitor titled *Illustrations of Masonry,* in London, England. After Preston, numerous monitors were published, including several works in America by such Masons as Chapman, Sickels, Macoy and Atwood. A Masonic guide published for the Wisconsin lodges in 1889 contained the poem, "The Level and the Square," authored by Rob Morris, that may be the most widely recognized Masonic poem. A most often repeated stanza is:

> "Hands round, ye faithful craftsmen, in the
> bright, fraternal chain;
> We part upon the Square below to meet in
> Heaven again.
> Oh, what words of previous meanings those
> words Masonic are—
> We meet upon the Level and we part upon the
> Square."

Of special note is a publication by George Smith in 1783 entitled *The Use And Abuse Of Freemasonry.* Prior to this date only handbooks or monitors sanctioned by a Grand Lodge had been published. Smith ignored the Grand Lodge, defied possible sanctions or repercussions, and published his book. From that date forward,

Freemasons were free to publish craft-related material without prior censorship.

Since this period literally thousands of books and pamphlets have been written about Freemasonry. Trivia buffs will recall that the first use of the term "freemason" appeared in two books that were published in 1563, one entitled: *The First and Chief Grounds...of Architecture* and the second, *A Book In English Metre, Of The Great Merchant Man Called Dives Pragmatius.*

Finally, no mention of Masonic literature should be made without touching upon a tangent field: musical compositions. In addition to Mozart's "Magic Flute," previously mentioned, other famous composers writing music specifically for or about Freemasonry include Goethe, Robert Burns, Franz List and Jean Sibelius.

The works and authors mentioned thus far represent only a partial list. It is impossible, for a work of this nature, to develop an exhaustive encyclopedia of all related literature.

In summation, what does this compilation of names, works, and dates actually signify to the contemporary Freemason? First, it reveals that Freemasnry was an accepted and acknowledged part of society by the 17th Century. Historians place great importance on being able to trace the actual, not the mythological, history of an event, person or organization. Written history, if accurately recorded, in the form of books, poetry, or music, tends to anchor facts, as opposed to oral history which can become more easily distorted, embellished and altered. A great deal of Masonic history can be garnered from a careful examination of the written word.

The use of references to Freemasonry also indicates the position of the Craft in society. A minor inconsequential organization could hardly be expected to generate the type of response and recognition the author hoped to elicit with their allusion to Masonic symbols and rituals. The fact that Russians, Englishmen, Frenchmen, Germans and Americans, among others, used Freemasonry in their works clearly illustrates the spread of the Institution across national borders. Freemasonry, then, was an accepted part of society and its outward, and perhaps supposedly esoteric, trappings recognized by the general populace.

From both of the aforementioned points a third factor can be deduced. By the 20th Century the parochial mysticism and secrecy of Freemasonry had evaporated. The pedagogical domain of the Craft had been invaded as education became more widely available to the masses and exposes of the Institution, *a la* Prichard, were widely published. The Philosopher's Stone remained undiscovered. Operative masonry gave way to speculative and esoteria was minimized. In short, moral humanitarian factors began displacing the vocational pragmatism. Freemasonry became available to the enlightened of all vocational pursuits. Once the Craft entered this new phase of evolution it entered the public domain and became the object of careful scrutiny. From this scrutiny came reference points that various authors utilized in order to elaborate upon themes they developed.

In truth, even the most diligent scholar and careful author could only speak to the aura surrounding the Institution. The emotional ties associated with the Craft could be captured, in part, by a Kipling or Burns, but the moral thread so carefully woven through the Fraternity, tied by ritual and participation, could not be experienced vicariously. Literature could bring the spirit of the Institution to the uninitiated, but only a superficial knowledge could be gained. In essence, the richness of Freemasonry's literary references can be fully understood by Freemasons themselves. What if Sherlock Holmes could identify an arc-and-compass breastpin? The importance of the passage focuses not on the pin, to the Initiated, but rather to the symbolism of the pin. The greater implications, to the profane, went unnoticed. The Prince of India observed the working tools of a Master Mason, but no explanation of their functional utility was forthcoming. Kipling could speak of a Lodge of Instruction or a nebulous "Third Degree," but what understanding could the non-Mason bring to these terms.

The institution of Freemasonry has been experiencing increasing visibility in all forms of written expression since the 18th Century. Most references, however, speak to the outward aspects of the Craft. Nonetheless, the simple fact that respected authors and composers have used the Craft to illustrate their efforts speaks to the universal appeal of this particular system of morality veiled in allegory and illustrated by symbols.

Yet, one final aspect of Freemasonry and literature remains to be discussed. In the aforestated examples, Freemasonry was the object of the authors' works, but the fact remains that Freemasonry drew much of its ritual related allegories from one of the greatest literary works of the past two millennium: the Bible.

Freemasonry is not a religion, but because of the inherent theistic philosphy of the Craft it is easy to see how religion and ritual became inextricably interwoven. In the Western World, Christianity and Judaism are dominant faiths, and it is understandable to justify the use of the Bible: Old and New Testaments, by Freemasonry. It must be recognized that the Bible can be interpreted as historical literature through which God is proclaimed as the chief focus of history Who alone gives history its meaning.

It is partially through the historical Bible that Freemasonry draws its teachings. Allegory and metaphor are indispensable aspects of Freemasonry and they appear frequently in the Old and New Testaments. Thomas Aquinas once asked whether it was proper for the Holy Scripture to use metaphors. He affirmed that many people object to metaphors and allegories because they believed divine truths should not, indeed cannot, be put forward by comparing them to things on earth. Aquinas stated: "...it is natural to man to attain to intellectual truths through sensible objects, because all our knowledge originates from sense. Hence, in Holy Writ, spiritual truths are fittingly taught under the likeness of material things." Aquinas postulates that allegories and metaphors are not only acceptable, but desirable because they communicate spiritual truths in understandable language. Freemasonry borrows heavily from the Writ.

In Freemasonry, for example, use is made of a sword pointing toward a naked heart. In Luke, 2, 33-35 the following passage keys the Masonic message:

> "And Joseph and his mother marvelled at those things which were spoken of him. And Simeon blessed them, and said unto Mary his mother, Behold this child is set for the fall and rising again of many in Israel; and for a sign which shall be spoken against; (Yea, a sword shall pierce through thy own soul also,) that the thoughts of many hearts may be revealed."

Many Masonic monitors speak to biblically related allegories, this paper will borrow from M. L. Young's 1889 *Masonic Guide or Monitorial Instruction* that was compiled for the Grand Lodge of Wisconsin. To this point Young wrote, "the sword pointing to a naked heart reminds that justice will sooner or later overtake us; and although our thoughts, words and actions may be hidden from the eyes of man..." they cannot be hidden from the Supreme Architect of the Universe.

Similarly, both the Bible and Freemasonry make use of the lamb and lambskin. Biblically, one may turn to Peter, I, 17-19 and Proverbs 27:25-26. (RSV) The former stating:

> "And if ye call on the Father, who without respect of persons judgeth according to every man's work, pass the time of your sojourning here in fear. Forasmuch as ye know that ye were not redeemed with corruptible things, as silver and gold, from your vain conversation received by tradition from your fathers; But with the precious blood of Christ, as of a lamb without blemish and without spot:"

The latter, more succinctly:

> "The hay appeareth, and the tender grass sheweth itself, and herbs of the mountains are gathered. The lambs are thy clothing, and the goats are the price of the field."

As might be discerned from the aforementioned citations, the lambskin is, for the Freemason, an emblem of innocence. To cite a recent Washington monitor: "The lamb has in all ages been deemed an emblem of innocence...by the lambskin the Mason is reminded of the purity of life and rectitude of conduct which is so essentially necessary to his gaining admission into the Celestial Lodge above..."

The interchange between the Bible, as literature, and Freemasonry, as ritual, is readily apparent. Without belaboring this point, the relationship can be further clarified by examing both the Biblical and Masonic use of the terms "the All-Seeing Eye" and the "Anchor and Ark."

In Hebrews 6:19 and 11:7 the anchor and ark are mentioned in the following context:

> "Which hope we have as as anchor of the soul, both sure and stedfast, and which entereth into that within the veil;"

and

> "By faith Noah, being warned of God of things not seen as yet, moved with fear, prepared an ark to the saving of his house; by the which he condemned the world, and became heir of the righteousness which is by faith."

To the Freemason the anchor and ark, to cite Young again are "emblems of a well-grounded hope and a well-spent life. They are emblematical of that divine ark which safely wafts us over this tempestuous sea of troubles, and that anchor which shall safely moor us in a peaceful harbor, where the wicked cease from troubling and the weary shall find rest."

The "All-Seeing-Eye," biblically, is well incorporated into the Masonic Fraternity's philosophy. From the Book of Psalms it is told:

> "The Lord is in his Holy Temple,
> The Lord's throne is in heaven.
> His eye is upon mankind, he takes their measure at a glance,
> The Lord weighs just and unjust."
>
> and
>
> "The Lord's eyes are turned toward those who fear him,
> toward those who hope for his unfailing love."

In the Masonic ritual it is revealed that "Whom the sun, moon, and stars obey, and under whose watchful care even comets perform their stupendous revolutions, pervades the inmost recesses of the human heart, and will reward us according to our merits."

Similar correlations can be drawn between other terms common to both works, such as the use of incense, the sycthe and acacia.

In retrospect, the influence of Freemasonry upon literature and the concomitant influence of the Holy Writ on the Craft dramatizes Matthew Arnold's century-old evaluation that "For the creation of a master-work of literature two powers must concur, the power of the man and the power of the moment." To this might be added: "the power of the movement." In this case: Freemasonry. If it is indeed true that in good writing words become one with things, good Masonic literature must become one with the Craft's philosophy. Not the outward crass allusions, but instead the inner-theological implications of Masonic thought.

Aesop, in his *Fables*, six centuries before Christ's birth, affirmed "Words may be deeds." For the Freemason to deny Aesop's contention is to ignore the impact of Masonry's literature, influence on literature and, above all else, Masonry's ritual dependence on the Holy Writ. Words may be deeds. *Masonically, our words must be translated into deeds.*

XIII

Freemasonry and the Future: Planning for Progress

As Freemasons, or members of concordant Masonic bodies, we are well aware of the intricate history of our Craft. We are justly proud of our heritage and find both satisfaction and comfort in our historical antecedents. This satisfaction, however, should not be mistaken as blind allegiance or misplaced perspective because, as Cicero stated, "To be ignorant of what occurred before you were born is to remain always a child." There comes a time, however, when the past, no matter how glorious and sublime, must be understood by this time-frame and used as a base for the present and a catalyst for the future. No viable organization can use the glories of the past to serve as an apologia for the future.

The future. What does it hold for Freemasonry? As contemporary Freemasons entering the last quarter of the 20th Century, are we assuming our historical and contemporary activities will prove sufficiently viable as to ensure the Craft's continued success?

Perhaps Eric Hoffer best capsulized this writer's sentiments when he wrote: "The only way to predict the future is to have the power to shape the future." Those interested in shaping Freemasonry in the 21st Century must start planning for that period today.

The rapid development of the academic discipline of futurism as a means of providing planners with resources for coping with the future can prove useful to Freemasons throughout the world. No organization, including Freemasonry, mindful of the future can neglect the process of planning, more specifically, acquiring the hard data required to construct a statistical profile of its membership. Without knowing the status of Freemasonry: where it has been and where it is now, how can meaningful decisions be made about planning the future?

Planning, to this writer, is used in a broad sense to encompass a rational means-end assessment of resources and objectives of Freemasonry for Freemasonry. Emphasis must be placed upon assessing the contemporary status of our Craft in order to gain the data required for developing long-range plans for future development. Management decisions cannot be made in a void, or crisis management, that is, responding to each new crisis in a knee-jerk manner.

Planning for the future is not simply desirable, it is an absolute necessity. Only the naive can assume that lofty moral virtues shall prove to be the sole effective shield that will ensure Freemasonry's position in the 21st Century. Although moral virtues are always desired assets, and to the Freemason indispensable factors in the Craft's structure and purpose, they must be coupled with effective short and long-range planning. At no time in its long and glorious history has Freemasonry required stronger leadership and long-range planning than it does at this moment. Although total membership figures for American Masonry appears static, or even reflective of modest increases or losses in some jurisdictions, a serious problem lurks, ready to appear at the turn of the century.

This writer has been concerned about the lack of hard data available for the effective administration/management of our Institution. This lack of data presents a potentially disastrous situation since it is difficult, if not impossible, to imagine how any institutions, including Freemasonry, can effectively plan for the future without understanding the *status quo*.

The *status quo* seems to have an ideological appeal to most administrators and, as a result, the vast majority of those in administration or managerial positions tend to be conservative. Organizations need stability in order to assure predictability of behavior and by creating a network of codes, rules and regulations; organizational activities are forced into certain patterns that assure a high degree of conformity. This entire process encourages conservatism and places the innovator at a severe disadvantage. As long as the organization meets its expected goals and objectives this process has merit and is justifiable. The problem arises, however, when there is dysfunctionality; that is, a gap appears between desired outcomes and actual outcomes. This creates a situation where an institution cannot respond in a manner that ensures its continued viability.

The individual Blue Lodge officers and Grand Lodge officers of Freemasonry are administrator-managers. In order to be effective they must possess the salient data that will enable them to make intelligent, logical decisions for the future. They must understand the *status quo*, add to that a relevant historical perspective and then develop long-range plans. This will take an organized effort which, logically, will be a structured effort. This structure can be derived from the grass-roots membership up to the Grand Lodge hierarchy or be developed at the managerial level and imposed downward *ipso facto*. Using the first approach there will be greater individual involvement, but it will prove to be a time-consuming process and possibly ill-defined. A management developed system will be quicker, but lack the grass-roots involvement. In this particular case, what is needed is a collection of data; the start of a data bank if you please, that will enable Masonic managers to develop a statistical profile of the Institution's membership. This data bank must be collected lodge by lodge, jurisdiction by jurisdiction and then assembled on a national basis. Whereas a unilateral code revision made at the Grand Lodge level would be unacceptable to a jurisdiction's Masonic population, this type of information gathering does not appear to infringe upon the rights and sovereignty of individual Masons or lodges.

Whatever process is utilized, some form of action must be taken. Administrative "slumber" will result in no activity and a lack of activity will allow the *status quo* to continue. In essence, the *status quo* allows unplanned change and unplanned change puts the future of Freemasonry on the fickle winds of fate. This cannot be allowed to happen.

This writer believes that Freemasonry needs to assemble salient statistic data far exceeding simple membership rosters and develop a complete profile of our populations. For example, after the 1977 Washington State Grand Lodge, this writer distributed 500 questionnaires to Masons from Alaska and Washington. Although the sample was undoubtably skewed by the qualification of the men attending a Grand Lodge, even a cursory examination of the responses should cause alarm. It is readily apparent that this regional population is heavily dominated by the upper age groups. Specifically, the average year a man joined the Masons was 1957 and at that time the average age was 35 years. It takes no mathematical genius to extrapolate the conclusion that by the year 1990, unless a radical change in the age structure is effected, the "average" Mason in this region will be 78 years. (Assuming, of course, that death does not intervene.) The obvious questions that spring to mind are: "Where are the younger men," and "Why are not more younger men Freemasons?" These questions have to be answered.

Detailed analysis is needed *vis-a-vis* why men join Masonry, why they attend lodge, why they do not attend lodge, what is the agestructure, what interests men about Masonry, what does not. Ancient traditions must be examined to ascertain the validity in contemporary society. Before such detailed analysis can occur, however, relevant data must be collected.

Unfortunately, Freemasonry's practices, rituals, history and future are too often defended by emotion and not by reason. This is not to belittle emotion, it is a God-given response that is extremely valuable in selected occasions. But, Freemasonry has nothing to fear from dispassionate analysis and there are pronounced hazards for the adminstrator/manager who works on the basis of his heart and not his head. Aside from the fact that feelings become easily bruised, much of what is decided by emotion has a short life span and once a crisis has passed, the emotions subside, but the causes of the problem will occur and re-occur with increasing regularity. Any administrator (lodge officers included) who attempts to become an "altar boy" and please everyone on every issue is doomed to failure. To ignore reality because fantasy is more comforting is suicidal. Decisions for the good of the Order must be effected and it is inevitable that some decisions will not be popular. This is regrettable, but unavoidable. The saving grace, of course, is that decisions made on the basis of facts are usually more accurate than those made on the basis of personal opinion.

It is of utmost importance to place the assembled data concerning Freemasonry in juxtaposition with data concerning the larger American society. For example, the fact that the Masonic population is aging gains considerable importance when one realizes that demographers predict a leveling or even a decrease in the number of live births in many sections of the United States. (Predicted increases will occur amongst non-whites and the female populations.) While high school graduates increased in the 1970's, the elementary and secondary school enrollment has gradually decreased since 1970. And it is estimated that there will be a gradual decline in high school graduates through the 1980's. Population growth will come from people living longer and not necessarily from new births.

Much the same as loggers cannot labor without trees or accountants cannot work without numbers, Freemasons cannot labor without new men. By drawing together the major dimensions of analysis needed to complete a valid statistical profile of Freemasonry, a model or paradigm can be constructed. This paradigm will prove useful in understanding the dynamics of Freemasonry and age-group fluctuations can be studied, codified and even predicted. Once the appropriate data-bank has been assembled, Masonic leaders can manage the Craft's future and not find themselves being surprised and managed by unsuspected events.

Whenever one comes to grips with problems dealing with Freemasonry it is an amusing diversion to imagine what Carl Claudy's "The Old Past Master" might have advised the myriad "Very New Masons" who always posed such poignant queries. This writer does not believe the Old Past Master would have been so very glib when faced with this perplexity.

Is this situtation really as serious as it appears at first blush? In the mid-1500's there was a religious sect in the Balkans called the Bogomils. They were a morally chaste group, believed in God, high moral values and regular worship. They did not solict membership; a prospective member had to seek them out. They were a wonderful group. There are no Bogomils today.

It is this writer's opinion that the greatest challenge to Freemasonry's existence did not occur with Pope Clement in the 18th Century, or with the William Morgan Affair or with any historical event that, in retrospect, has failed to crumble the foundation of our Craft. The greatest challenge is yet to come and it will be presented not on the stage of public scrutiny, but will spread from within our ranks. *Apathy coupled with inaction may erode away the foundation of Freemasonry as age robs us of our youth.* A viable Craft must enjoy a healthy balance of old and young. It is true what the Bible tells us in Job 12:12: "With the ancient is wisdom; and in length of days understanding," but it is equally true what Somerset Maugham stated in *The Summing Up:* "For the complete life, the perfect pattern includes old age as well as youth and maturity." Freemasonry needs a mixture of ages: young and old in order to pass its wisdoms and teachings on from one generation to the next. Are we approaching an era where this "passing of the torch" is endangered? Preliminary evidence tends to indicate a crisis is slowly developing. This writer sincerely hopes he is wrong, but only further research and then accurate long-range planning can bridge this chasm.

Aside from the statistical evidence that remains hidden, waiting for the research instrument to bring it to the light of day, other, more subtle indications of Freemasonry's status is available. Ask yourself, who are the great Masonic leaders of today? Who have we produced to take the place of Albert Pike, Carl Claudy, Joseph Fort Newton, Roscoe Pound, Robert Freke Gould and Albert Mackey? Rudyard Kipling, Robert Burns and even Conan Doyle venerated Freemasonry with the written word in popular literature Where are such authors and playwrights today? Masonic men of repute whose names rise to most Masons' lips are usually political figures or officials of governing Masonic bodies.

It is this writer's contention that a crisis situation is at hand. So, as we sow, so shall we reap. The fruits of the 21st Century must be planted now or the blossoms will never appear and our trees will grow barren, whither and die. The question focuses on whether Freemasonry is to remain a viable productive member of human society or shall it begin to fade into the annals of man as a glorious, but extinct, footnote. The problem is not today, but rather 1990, 1995 and the year 2,000. It is obviously terribly overdramatic to ring the bells of alarm as the result of a survey that reached only 150 men. But what if the limited results are an accurate barometer of the larger body? More research is needed, good research, directed research that will give the Freemasons of each and every jurisdiction the information they need to guide their own directions.

XIV
Masons as Managers

Perhaps another aspect of the continuing dilemma focusing on Masonry, and the future, centers on the role to be assumed by the lodge's chief executive officer.

Ask most Freemasons what makes a good Master of a lodge and the traditional answer will probably focus on some aspect of ritualization. A man that can recite verbatim Masonic rituals from the first word of the first degree to the last word of the third degree, including all posting lectures, has historically been regarded as an ideal candidate for assuming the chair in the East. When pressed further on this topic, it is affirmed that a good Master must also be able to handle the business of the lodge, organize and staff committees, select projects, allocate resources and be responsible for the lodge's general welfare. But, these thoughts are usually secondary to the ritualistic aspects of a Master's responsibilities.

Carl H. Claudy, a well-known Masonic author, placed a Master's tasks in perspective in his tiny volume entitled *The Master's Book*. Claudy, in Chapter 3, lists 14 functions of a Master and *only one* focuses on degree work. Reference to degree work is made only after 6 other functions are enumerated and commented upon. Obviously, attention to degree work is vitally important, but it is not the end-all that many Freemasons erroneously believe to be the sole requisite to qualify a man for the East. Lodge *management* is the prime function of a lodge's chief executive: to make sure what needs to get accomplished does indeed get accomplished. In order to handle this particular duty, a Master must use good judgment and good judgment represents the not too simple task of matching known facts to values: in this case Masonic morality. How does this topic affect the future, or more importantly, the future of Freemasonry?

Phenomena of lasting import are sweeping through Freemasonic lodges throughout the land. Radio, television, theaters, sports events, and other activities that reach out to the masses are now competing very effectively with the lodgeroom for an individual's time and energies. Times have changed and the contemporary Freemason is bombarded by external stimuli that never existed in our grandfathers' time. A lodge's chief executive must be able to marshall his assembled resources (the affiliated Masonic brethren) in order to produce a functional/salable product. Does it sound mercenary and undignified to speak of Freemasonry in such "Madison Avenue" type terms? Perhaps, but Freemasonry's historical existence and future growth, all due to esoteric knowledges, appear severely limited and perhaps in doubt.

The Master of the lodge must no longer content himself with rotely memorizing paragraph after paragraph of various rituals and then spewing them forth, upon cue, even in the most eloquent and heartfelt manner. The Master, or chief lodge executive, must find symmetry between what is occurring in the lodgeroom and what is happening in the world without. Although often administering several functions at once, he cannot allow himself to become distracted from his previously determined list of priorities as related to his lodge's future. The harried Master who cannot distinguish between the incidental and the essential will undoubtedly perform in an impulsive and often premature manner. Whereas, in periods of growth this manner of management, while not excusable, did not appear to harm the Craft, it can no longer be deemed acceptable. It has been said that the functional specialization of activities in an organization inhibits balanced and comprehensive management. Perhaps in the profit-oriented business world this is true, but such functional specialization in a volunteer organization such as Freemasonry is not only desirable, but mandatory. Delegation of authority not only relieves the Master of tedious details of time-consuming chores, it also involves countless members in lodge activities. Such involvement breeds interest and interest breeds activity. The more involved the membership becomes on a lodge-wide basis, the greater the development of the human resources to be made available for the future.

Claudy does not disagree with this premise. He states that the Master's first duty to his lodge is to lead it to success and prosperity. Included in this leadership he enumerates such activities as the publishing of a monthly bulletin, attention to lodge finances, planning lodge entertainment, promoting promptness, attending the sick, as well as overseeing degree work. Claudy states:

> "The ideal Master is primarily concerned with policies rather than details, and delegates the latter to carefully chosen committees."

What prevents the ideal lodge administrator from freeing himself from the cumbersome daily details and focusing directly on long-range planning? First and foremost, Freemasonry is not devoid of intra-fraternal factioning. Although theoretically bound by solemn oaths that prompt cooperation, cliques do form and inhibit harmony. Secondly, although the Master is the chief executive of the lodge, as with all administrators/managers, honest differences of opinion exist and can cause disharmony. Past Masters can prove to be extremely valuable to a lodge, but if they so desire, one or more can prove to be a devisive force within the lodge. Why? It is natural for any administrator to have a strong desire to continue that with which he is familiar. When a new adminstration assumes office, the Master has two prime courses of action: (1) he can maintain the *status quo* or (2) he can establish a new plan-of-action that deviates from previously established patterns. Organizational hangover from previous lodge adminstrations can either promote or hinder the new Master's task.

What does this discourse on lodge management have to do with the broader topic of Freemasonry and the future? A great deal, actually, for as the old bromide relates: a journey of a thousand miles commences with the first step. Before Freemasons can begin to plot the course of future global maneuvers, the first step is to make sure that the basic cornerstone of the Institution, the Mother lodge, is secure and progressive, looking to the future and not squabbling amongst themselves. Secondly, at some point in time Freemasonry will have to examine the ancient traditions and usages passed down to the modern lodges from our historical brethren and determine which are still valid and, possibly, which might be counterproductive in contemporary society. The simple fact of existence is that nothing stays static. Everything changes.

There is nothing written in Masonic codes and regulations that is etched in stone, save the concepts embodied in the Ancient Landmarks, that cannot be altered. This fact is readily apparent when one examines the differences that exist even between adjoining Grand Jurisdictions. Freemasonry has always been a growth Institution, that is, it has altered and adapted to fit contemporary societies. A case-in-point focuses on the step from Operative Masonry to Speculative Masonry. No doubt certain traditionalists of that era saw this transformation as the beginning of the end of their Craft. It probably caused a great gnashing of the teeth and beating of the breasts. Yet, if this step had not been taken not only would there be very few Speculative Masons alive to read these words, these words would not have been written. No one would have had the advantage of seeing Masonic Light as we know it today. Are there any Master Masons living today who feel that change, so profound in nature, was not beneficial to mankind in general and themselves, specifically? Change will come, the quesiton is: do we want to control change or do we want to be controlled by it?

This writer believes Freemasonry is at a juncture in time of equal proportions to the shift from Operative to Speculative Masonry. Today, our Craft grows old, the young men no longer flock to our doors as they have in the past decades. Some caution that the pendulum swings to and fro, and if we wait our numbers will grow again. But will they? The human desire for comfort through familiarity appears to have snared Masonic lodges into using certain traditional practices as an apology for the state of the Craft. This prevents a useful examination of what is productive and what is dysfunctional and counterproductive.

Understanding new information and new knowledges is an extension of all previous cognitive knowledge stored within an individual. The ability to assimilate new knowledges and use such for future reference aids in the development of functional systems. If a system is not functional, it exists only for cosmetic purposes and lacks power or direction. Freemasonry has historically been a functional organization both as an operative craft and as a speculative institution. This change affected only the form of functionality. Contemporary Masons are antecedents of this altered functionality and, as such, we remain in varying degrees, functional today. But what of the future? Can Freemasonry remain a viable functional system after the turn of the 21st Century?

No one can accurately predict the future, but this does not mean we should not plan for the future. We can assess existing knowledge, gather new data, and extrapolate logical hypotheses on the basis of assembled information. To fail to do so is to blindly trust in the winds of fate. Freemasonry must, to ensure its future, focus on the future. Lodge management must exceed dramatized rituals. Current practices must be scrutinized for their functionality in contemporary society. Masonic history must continue to be studied, but not used as a crutch to justify a hypothesized future. Each Freemason actively involved in Masonic activities must remember that simple children's saying: "If it is to be, it is up to me."

It is possible that after all data are assembled and interpreted that we will find Freemasonry requires no alterations. Or, we may find changes are indeed required. We will never know anything, however, until research is effected. Research for the future, so Freemasonry can shape its future and not be subjected to the pangs of unexpected crisis. Planned change, that is the key to Freemasonry's future. No operative mason would approach a task without his square, level and plumb. No Speculative Mason should be any less prepared.

So mote it be.

Footnotes and References

[1] Albert Pike. *Morals and Dogma of the Ancient and Accepted Scottish Rite.* L.H. Jenkins, Inc., 1871 (p. iii)

[2] Norman MacKenzie, ed. *Secret Societies*, Holt, Rhinehart and Winston. 1967 (p. 12)

[3] J.M. Roberts. *The Mythology of Secret Societies*, Seeker and Warburg. 1972. (p. 60)

[4] MacKenzie. *op. cit.*, (p. 17)

[5] William Preston. *Illustrations of Masonry.* J.W. Brennan. 1855. (p. 4)

[6] MacKenzie. *Op. cit.* (p. 17)

[7] Pike. *Op. cit.* (p. 17)

[8] Daniel Sickels. *General Ahiman Rezon and Freemason's Guide.* Masonic Publishing Company. New York. 1865. (p. 14)

[9] MacKenzie. *Op. cit.* (p. 17.)

[10] Pike. *Op. cit.* Note—Pike is adamant concerning this particular premise. In the appendix to his work he lists no fewer than 150 references to this one topic.

[11] Joseph Fort Newton as cited by the Masonic Service Association.

[12] Robert Gould. *History of Freemasonry.* The Claxton Publishing Company. London. 1951. Volume II.

[13] Obadiah Harris. *Unitive Spirituality.* J.F. Rowny Press. Santa Barbara. 1966 (p. 79)

[14] MacKenzie. *op. cit.*

[15] Henry Stillson, Ed., *History of Freemasonry.* The Fraternity Publishing Company. London. 1890. (p. 90)

[16] *Ibid.*

[17] Gould. *Op. Cit.* p. 90

[18] *Ibid.* P. 115

[19] Henry Wilson Coil. *Freemasonry Through Six Centuries.* 2 Vols. Macoy Publishing & Masonic Supply Co., Inc., Richmond, Virginia. 1967. (p. 108)

[20] *Mackey's Revised Encyclopedia of Freemasonry.* Revised by Robert Ingham Clegg. Macoy Publishing & Masonic Supply Co., Inc. 1966 (p. 689)

[21] *Ibid.* pp. 690-691.

[22] Coil. *Op. Cit.* p. 194

[23] Mackey. *op. cit.* p. 907.

[24] *Ibid.* p. 941.

[25] Rudyard Kipling. *Debits and Credits.* MacMillian & Co., London. 1965. (p. 55)

[26] *Ibid.* p. 56.

[27] Rudyard Kipling's Autobiography.

[28] Edgar Allen Poe. "The Cask of Amontillado," *The Complete Works of Edgar Allen Poe.* edited by J.A. Harrison. AMS Press. New York, New York. 1965. (pp. 171-2)

[29] A. Conan Doyle. "The Red Headed League." *Adventures of Sherlock Holmes.* Harper and Brothers. New York and London. 1917. (p. 31)

[30] J.M. Roberts. *Op. Cit.* p. 62.